The Physics of Supernovae and Their Mathematical Models

The Physics of Supernovae and Their Mathematical Models

Alexey G. Aksenov
Valery M. Chechetkin

Russian Academy of Sciences, Russia

NEW JERSEY • LONDON • SINGAPORE • BEIJING • SHANGHAI • HONG KONG • TAIPEI • CHENNAI • TOKYO

Published by

World Scientific Publishing Co. Pte. Ltd.
5 Toh Tuck Link, Singapore 596224
USA office: 27 Warren Street, Suite 401-402, Hackensack, NJ 07601
UK office: 57 Shelton Street, Covent Garden, London WC2H 9HE

Library of Congress Cataloging-in-Publication Data
Names: Aksenov, Alexey G., author. | Chechetkin, V. M. (Valeriĭ Mikhaĭlovich), author.
Title: The physics of supernovae and their mathematical models / Alexey G. Aksenov, Valery M. Chechetkin, Russian Academy of Sciences, Russia.
Description: New Jersey : World Scientific, [2024] | Includes bibliographical references and index.
Identifiers: LCCN 2023047037 | ISBN 9789811285097 (hardcover) | ISBN 9789811285103 (ebook for institutions) | ISBN 9789811285110 (ebook for individuals)
Subjects: LCSH: Supernovae. | Supernovae--Mathematical models.
Classification: LCC QB843.S95 .A25 2024 | DDC 523.8/4465--dc23/eng/20231116
LC record available at https://lccn.loc.gov/2023047037

British Library Cataloguing-in-Publication Data
A catalogue record for this book is available from the British Library.

For any available supplementary material, please visit
https://www.worldscientific.com/worldscibooks/10.1142/13644#t=suppl

Desk Editors: Nimal Koliyat/Ana Ovey/Shi Ying Koe

Typeset by Stallion Press
Email: enquiries@stallionpress.com

Printed in Singapore

About the Authors

Alexey G. Aksenov was born on September 23, 1966. He is a Senior Researcher at the Institute for Computer-Aided Design, Russian Academy of Sciences (ICAD RAS) in Moscow. He graduated from the Moscow State Engineering Physics Institute (MEPhI, Technical University) and holds a PhD in astrophysics from the Space Research Institute of the Russian Academy of Sciences. He is the author of one monograph and more than 40 refereed publications on different topics in astrophysics and plasma physics, gravitational collapse, neutrino transport, inertial confinement fusion, numerical solution of kinetic Boltzmann equations, and hydrodynamic simulations.

Valery M. Chechetkin was born on March 23, 1941, in Moscow. In 1964, he graduated from the Moscow Institute of Physics and Technology. Since 1968, after completing his postgraduate studies at the Institute of Applied Mathematics of the USSR Academy of Sciences, he worked at M. V. Keldysh Institute of Applied Mathematics RAS in positions from Junior Researcher to Chief Researcher, Dr. Phys.-Math. Sci. (1984), Professor (2000), Professor of MEPhI, and he is also the author of over 400 publications. He has won several state awards, including Honored Scientist (2006) and G. A. Gamow (2004).

He was a Member of IAU from Russia and also Chevalier of Ordre des Palmes academiques (2017). V. M. Chechetkin is an expert in the field of theoretical astrophysics and fundamental physics, and also one of the founders of physical and mathematical modeling for astrophysical processes and phenomena.

The following achievements by V. M. Chechetkin can be highlighted:

- Chechetkin obtained the structure of the explosion in Type I SN, which, when using these supernovae as standard candles, makes it possible to draw conclusions about dark energy in the universe.
- He proposed a new mechanism for the formation of heavy elements, which makes it possible to determine the age of a galaxy on the basis of the yield of uranium–thorium isotopes and to reveal the properties of the physical vacuum. This enabled in 1987 placing restrictions on dark matter and on the density of dark energy.
- He proposed the process of development of large-scale convection in a proto-neutron star, which leads to an increase in the average energy in neutrino radiation at early stages and accelerates the output of a neutrino pulse.
- He obtained restrictions on the primary content of antimatter in the universe and restrictions on possible new unstable particles based on the study of the kinetic processes of the interaction of antiprotons with helium-4 and the abundance of light elements. He also obtained restrictions on the concentration of primordial black holes.
- He obtained restrictions on the axion mass from astrophysical estimates.
- He discovered the nuclear meson mechanism of the formation of neutrino and gamma radiation from relativistic shock waves and during the accretion of matter onto the surface of a neutron star.
- For the first time in the world, three-dimensional unsteady calculations of MHD flows under astrophysical conditions of accretion disks were performed, the flow structure was constructed, and the mechanisms of matter outflow in jets were proposed. The large-scale structure of magneto-rotational instability was investigated under astrophysical conditions.
- He investigated the flow of matter in close binary stellar systems. The results obtained in a new way represent the distribution of

matter in binary systems and highlight new hydrodynamic structures (for example, the existence of a common shell and the emergence of detached shock waves from rotating stars in it). When interpreting observations of close binary stellar systems, proved the incorrectness of the "hot spot" model, which contradicts the laws of gravitational hydrodynamics.

- He proposed a physical model of the formation of large organized structures in turbulence for a free shear flow and discovered the mechanism of the formation of a cascade due to the secondary instability of large structures in a three-dimensional flow.
- He proposed a mechanism for the development of large-scale turbulent structures in accretion disks, leading to a redistribution of angular momentum without heating the matter.

Contents

Introduction

The aim of this book is the generalization of long-time experience of authors in theoretical astrophysics of compact objects. One author (V. M. Chechetkin) started his work in the theoretical astrophysics group by Ya. B. Zeldovich in Keldysh Institute of Applied Mathematics of Russian Academy of Science in 1969. The other author started his work in the laboratory of plasma physics and astrophysics by V. S. Imshennik in the Institute of Theoretical and Experimental Physics (ITEP) in 1989. ITEP, established on December 1, 1945, initially carrying the name "Laboratory №3", now is a part of the Kurchatov Institute. Both the authors can be attributed to the Soviet School of Supernova Physics, in particular, in the numerical simulations. For this reason, we emphasized old Soviet articles in this field.

Because one author (Prof. Chechetkin) has long-time experience with a course of lectures at Moscow Engineering Physics Institute (National Research Nuclear University MEPHI), and both the authors have active cooperations with foreign science centers in the formulation and the solving of astrophysical tasks, this book can be considered not only a collection of results, but it is also a practical tool for investigators and postgraduate students. For this reason, this book consists of three parts: Theoretical Foundations and Investigations, Numerical Methods, and Applications.

Part 1: Theoretical Foundations and Applications contains the a nonformal introduction into the basics of supernovae, Boltzmann kinetic equations — with details of two particles reaction rate calculations — and the transformation of Boltzmann kinetic equations into hydrodynamic elements of statistical physics. It also contains the equation of state for matter of high energy density, with details of calculations for thermodynamic parameters, weak interactions, reaction rate details, and thermonuclear burning.

Part 2: Numerical Methods collects not only the classification of equations of mathematical physics and generally accepted simulation methods but also the original approaches of the authors. Partially this part repeats some materials from recent book [309], but this part also includes some advantages of last years. One should be warned there is no universal set of numerical receipts applicable for the whole range of physical problems. Instead of some numerical algorithms that now make part of such computational programs as Mathematica or MATLAB [317], the army principle "do as I do" in computational physics is working partially. It has to be emphasized that basic ideas in computational physics come from theoretical and mathematical physics like Part 1. At least one author (A. G. Aksenov) found useful only one recommendation: to start creating the computation program from the mathematical formula of the developed method.

Part 3: Applications contains our original works in the regime of burning in degenerate carbon–oxygen cores, a neutrino transport in Type II supernovae, simulation of general relativity (GR) coalescence of neutron stars, aspherical nucleosynthesis in a core-collapse supernova, and thermalization in a pair of plasma winds from a compact strange star. Unfortunately, because of limited time, we did not include the review of the first Soviet publication about accretion discs, correlation between SN and gamma-ray bursts, and formation of elements beyond the iron peak (*rbc*-process).

The authors acknowledge colleagues of publications used in this book and also their teachers Ya. B. Zeldovich (for V. M. Chechetkin) and V. S. Imshennik (for A. G. Aksenov).

Part 1

Theoretical Foundations and Investigations

Chapter 1

Evidence of Supernovae

1.1 Supernova parameters

Supernovae are stars that finish their late evolution in a catastrophic explosive process. The term "supernova" was introduced and the difference between supernovae and novae in terms of their estimated explosive powers was described in Ref. [51]. The amount of energy emitted by the Sun over 10^7 years is emitted by a supernova over 25 days, while the power of a nova exceeds the solar power by only a few orders of magnitude (a factor of 20,000). The luminosity of a supernova at its maximum (lasting for several days) is comparable to the total *luminosity* of its host galaxy.

It was first hypothesized in Ref. [50] that supernova explosions should be accompanied by the formation of neutron stars; the neutron had been discovered just two years earlier. Before the discovery of the neutron, Landau [198] had considered atomic nuclei to be part of a giant nucleus (see also Ref. [322]). The explosion energy can be estimated in the following simple way. If a solar mass, $M_\odot = 2 \cdot 10^{33}$ g, is compressed to a radius of $R_{\mathbf{NS}} = 10$ km, a neutron star with a gravitational energy of $\sim 10^{54}$ erg (approximately $-GM/R_{\mathrm{NS}}$) will be formed, i.e., this implies an energy supply of several orders of magnitude for a Type II supernova. Thus, a nucleon with a mass of $m_n = 1.67 \cdot 10^{-24}$ g accounts for $\sim$100 MeV in the compression process, while less than 10 MeV/nucleon is released during thermonuclear burning in stars.

Observations of SNs have been carried out for several thousand years. The first physical–mathematical models were formulated and solved in the 1960s [53,98,100,109,171,174,177,235,236,287]. Supernovae in our galaxy are believed to explode no more often than once every 100 years. In the recent past, approximately 20 supernova per year were discovered using telescopes and photographic receivers. Now, more than 100 distant supernovae, for which spectra and light curves are available, are discovered each year using modern CCD arrays [62,175].

The type of a supernova is defined by the presence (or absence) of a hydrogen absorption line that can arise in the spectrum due to the presence of a cool envelope. This absorption line is present in the spectra of *Type II supernovae* and absent in the spectra of *Type I supernovae*. In addition, Type II supernovae normally contain compact remnants, although such a remnant was not found in the nearby supernova SN 1987A [54].

Another supernova model describing Type I supernova is related to the thermonuclear explosion of a degenerate carbon–oxygen core [37,83]. This model is relevant for stars in binary systems with mass transfer. In a main-sequence star with a bound mass of $M \lesssim 10\ M_{\odot}$, the carbon–oxygen envelope can explode, whereas the core collapses in a more massive star. During the burning, the carbon-burning deflagration regime can be transformed into a detonation regime due to the ignition of neutrinos and adiabatic heating, yielding an envelope kinetic energy of $\sim 10^{51}$ erg [53,138,178,181,183,328]. At the same time, a pulsed burning regime is realized in deflagration [53,178,179]. More modern studies of the thermonuclear burning regimes of Type I supernovae were performed in Ref. [172]. In non-degenerate matter, the pressure behind the detonation wave increases the temperature not enough to support the wave, and absorption in the envelope) were modeled (under specific assumptions) without solving the transport equation. The equation of state assumed an equilibrium mixture of nuclei with specified c, T, and Y_e, electron–positron pairs, and photons in equilibrium.

Moreover, different types of supernovae display different light curves. The light curves of Type II supernovae have flat "knees" due to the passage of the shock through the massive envelope; Type I supernovae explode rapidly, and their *light curves* do not have "knees".

However, SNs remain incompletely understood to the present day [38,62,175]. This is related in part to the complexity of the physical-mathematical models required. In addition, there are few or no direct observational data on the gravitational collapses of stellar cores — spectra for various types of neutrinos, data on the gravitational signal — that could be used to help identify suitable mechanisms for the explosion and distinguish key physical processes. In the meantime, a variety of ideas continue to be expressed about the initiation of the neutrino flux during the collapse associated with the thermonuclear explosion in the envelope of the pre-SN [100], including the suggestion of a phase transition to a quark star [102]. Even such models can be studied using the codes we have developed. The situation with regard to observational data that can be obtained has improved substantially, in connection with the construction of new neutrino observations that are more sensitive than those in existence during the explosion of SN 1987A [273] (by orders of magnitude), as well as the construction of new gravitational-wave detectors [149]. Therefore, the development of physical–mathematical models incorporating sufficiently complete physics remains of interest [223,246]. Over the last decade, convincing evidence for a connection of some types of gamma-ray bursts with SNs has been obtained, as well as models to explain this link [96,280].

Because of the differences in the timescales involved, analyses of the explosions of collapsing supernovae are usually divided into two sections: the gravitational collapse itself and computation of the photon light curves [175]. Light-curve computation is the most well-studied aspect of supernova theory: The availability of extensive observational data on light curves in the visible and X-ray and well-developed numerical models for radiative hydrodynamics makes it possible to determine the explosion energy at the center of the SN and the chemical composition of the pre-supernova. At the same time, in spite of the completeness of the data (the physical formulation of the problem), an explosion occurring during a gravitational collapse remains an unsolved problem. There exist a fairly large number of models for pre-supernovae obtained from evolutionary computations. At the end of its evolution, a massive star (10–25 $M_\odot$) uses up its store of nuclear fuel, and the iron core, with a mass of 1.2–2 $M_\odot$, begins to collapse due to *core neutronization* (low-mass cores) or *photodissociation of iron nuclei* at high temperatures (high-mass cores) [171,177]. This collapse is accompanied by energy losses to neutrino radiation. During the collapse, the density and temperature are enhanced, and various types of neutrinos carry away an energy comparable to the gravitational binding energy of the proto-neutron-star core of the star in its final state. To explain the SN explosion, we must understand the mechanism imparting an energy of $(1\ldots1.5)\cdot10^{51}$ erg from the flux of neutrinos with energy $(1\ldots3)\cdot10^{53}$ erg to the SN envelope [109]. Unfortunately, no self-consistent computations have demonstrated a means for transferring 1% of the emitted energy to the pre-SN envelope [189].

1.2 Historical supernovae

In fact, supernovae and novae are not new stars: These explosions occur on existing stars. Historically, however, stars that became visible during outburst had previously not been seen in the sky; these were even called guest stars. Historical supernovae in our galaxy (which could be seen with the unaided eye for months) were observed by Chinese, Japanese, and European astronomers: SN 185 (its remnant gives a strong X-ray image now), 837, SN 1006, SN 1054, SN 1181, SN 1408, SN 1572 (Tycho's supernova), and SN1604 (Kepler's supernova). The most brilliant supernova ever appeared on May 1, 1006. It was observed in China and in the Middle East and Europe. It was "bright enough to cast shadows on the ground at night, brighter than the quarter moon" [217]. Its remnant can be observed now as a radio image. Marshall [217] recounts amusing astrological prediction in connection with the supernova of 1006, as well as that of 185.

The most famous old supernova is that of 1054, recorded by the Chinese but not reported in Europe. Its remnant is the Crab Nebula. It differs fundamentally from the remnants of 185 and 1006, which show only as radiant shells, representing the shock wave that these supernovae have sent out into space. In the Crab Nebula, a whole volume is luminous. This is connected with the a neutron star (a pulsar) in the center, which emits electromagnetic radiation of all frequencies at regular intervals 30 pulses per second. The pulsar probably also emits electrons which irradiate the material of the remnant and make it emit visible polarized light. The remnants of 185 and 1006 have no pulsars in the center.

In November 17, 1572, the Danish astronomer Tycho Brahe discovered a "new star" in Cassiopeia, which is brighter than Venus. He determined its exact position on every clear night during the several months the star was visible and found that it did not change relative to the fixed stars. The remnant of the Tycho supernova has no pulsar in it.

Kepler, in October 1604, saw another supernova, less bright than Tycho's but remaining visible for a whole year. Its position in the sky was accurately described by the astronomers, and the X-ray emission from its remnant has been measured.

Another supernova exploded in our galaxy between 1650 and 1680, known as Cas A (Cas for Cassiopeia). Its remnant is a very strong radio source, but it was not reported by contemporary observers; it is described in detail by Marschall [217]. Some supernovae in other galaxies were observed between 1885 and 1930 [217].

The latest nearby *supernova SN 1987A* exploded in a nearby galaxy; this could also be seen with the unaided eye, although only by those with sharp vision (it was observed in the southern hemisphere).

Supernovae have played a particularly important *role in history and chronology.* Contemporary data on the sizes of supernova remnants, the supernova-envelope expansion velocities, and the distances to supernovae can be used to determine the ages of supernovae using the *Sedov solution* [275] for a strong explosion with the energy E_0 in the interstellar medium with the density ρ_0. In this case, the only dimensional combination which contains only length and time is the ratio of E_0 to ρ_0, with the dimensions $[E/\rho_0] = \mathrm{cm}^5 \cdot \mathrm{s}^{-2}$ [326]. Hence, the dimensionless quantity

$$\xi = r\left(\frac{\rho_0}{Et^2}\right)^{1/5} \tag{1.1}$$

can serve as the similarity variable. The shock front is defined by a given value of the independent variable ξ_0. The motion of the wave front $R(t)$ is governed by the relationship

$$R = \xi_0\left(\frac{\rho_0}{Et^2}\right)^{1/5} t^{2/5}. \tag{1.2}$$

The propagation velocity of the shock wave is

$$D = \frac{dR}{dt} = \frac{2}{5}\frac{R}{t} = xi_0\frac{2}{5}\left(\frac{E}{\rho_0}\right)^{1/5} t^{-3/5} = \frac{2}{5}\xi_0^{5/2}\left(\frac{E}{\rho_0}\right)^{1/2} R^{-3/2}. \tag{1.3}$$

In addition, the *ages of supernovae* can be independently derived from observations of the spin-down of pulsars in supernova remnants and the corresponding estimates of the pulsar ages. Comparison of these data with Chinese chronicles suggests that the uncertainties in estimating historical supernova ages are ±100 years [169], hindering further revision of the basis of chronology.

Zwicky and Baade began a *systematic study of supernovae.* Zwicky was a Swiss physicist, and Baade a astronomer; both worked in Caltech. Zwicky obtained one of the newly developed Schmidt telescopes, which can photograph a fairly large area of the sky. His first Schmidt had a diameter of 18 inches. Within a year, we discovered three supernovae, and he and his assistant, J. J. Johnson, found near 20. They are detected by comparing the picture of a galaxy at two different times: If the later picture shows a bright spot where the earlier one had none, this is likely to be a supernova. Minkowski (also of Caltech) measured the spectra of the discovered SNe. Together the astronomers found two types of SNe: Type II has strong lines of hydrogen and Type I has none. The first 12 of Zwicky's SNe were all Type I, but number 13, Discovered by Johnson, was Type II.

Supernovae in our galaxy are believed to explode no more often than once every 100 years. In the recent past, approximately 10–30 supernova per year were discovered using telescopes and photographic receivers. Now, more than 100 distant supernovae, for which spectra and light curves are available, are discovered each year using modern CCD arrays [175].

Chapter 2

Kinetic Boltzmann Equations, Collision Integral, and Hydrodynamics

The Boltzmann kinetic equation is a nonlinear integro-differential equation, and the unknown function in the equation is a probability density function in six-dimensional space of a particle position and momentum. We limit the consideration of one-particle distribution function (DF) for the gas or for every king of particles of a plasma $f(\mathbf{r}, \mathbf{p}, t)$. We are interesting a system from a large number of particles composed. The characteristic number of particle is the Avogadro number $N_{\mathrm{A}} = 6.022 \cdot 10^{23}\mathrm{mol}^{-1}$. So, it is possible to hope that it is not necessary to describe a system as an exact particle-to-particle interaction. We expect it is possible to describe the system by a continuous one-particle distribution function $f(\mathbf{r}, \mathbf{p}, t)$ that is defined on the 6D phase space $(\mathbf{r}, \mathbf{p})$. It is expected to be true for two cases: a rarefied gas interacting mainly binary or an ideal plasma of charged particles interacting on a long distance but with a large number of particles in a Debye sphere with a radius $r_{\mathrm{D}} = \sqrt{\frac{4\pi n e^2}{k_{\mathrm{B}} T}}$. DF depends on seven variables: three space coordinates, three momentum components, and time. The one-particle distribution function is then defined such that the integral

$$N = \int f(\mathbf{r}, \mathbf{p}, t) d\mathbf{r} d\mathbf{p} \tag{2.1}$$

gives the total number of particles, and

$$n(\mathbf{r}, t) = \int f(\mathbf{r}, \mathbf{p}, t) d\mathbf{p} \tag{2.2}$$

is the concentration. $f(\mathbf{r}, \mathbf{p}, t) d\mathbf{r} d\mathbf{p}$ is an average number of particles having momenta in the range $(\mathbf{p}, \mathbf{p} + d\mathbf{p})$ and coordinates in the range $(\mathbf{x}, \mathbf{x} + d\mathbf{x})$ at the moment t.

2.1 One-particle distribution function

The basic function used in kinetic theory, for which master equations are formulated, is the *one particle distribution function.* In fact, all macroscopic information about the evolution of the system can be obtained from this function by suitable integration in the *phase space.*

Two approaches exist in the literature on relativistic kinetic theory regarding the definition of the *phase space* for one-particle distribution function. In the first approach, particle coordinate four-vector x^μ and particle momentum four-vector p^μ are considered as basic independent variables, and the one-particle distribution function $\mathcal{F}(x, p)$ is defined on the 8D *phase space* $\mathcal{M}^8$ with the help of particle four-current $j^\mu(x)$ as [309]

$$j^\mu(x) \equiv c \int \mathcal{F}(x, p) p^\mu d^4 p, \tag{2.3}$$

where c is the speed of light. For brevity, in what follows, we denote the coordinates in the *phase space* as $x \equiv x^\mu = (ct, \mathbf{x})$ and $p \equiv p^\mu = (p^0, \mathbf{p})$, where t is the time, $\mathbf{x}$ is the coordinate, and $\mathbf{p}$ is the momentum three-vectors, respectively. The distribution function defined in Eq. (2.3) is manifestly Lorentz invariant. However, this function, unlike the non-relativistic distribution function, does not have a meaning of the density of particles in some space.

In the second approach, the one particle distribution function $f(x^\mu, p^\mu)$ is defined on the 6D *phase space* $\mathcal{M}^6$. This function depends only on seven variables: three space coordinates, three momentum components, and time. This choice is evident since particle energy $E = cp^0$ depends on particle momentum, and p^0 satisfies the relativistic equation $p^0 = \sqrt{\mathbf{p}^2 + m^2 c^2}$, where m is the particle mass.

The one particle distribution function is then defined such that the integral

$$N \equiv \int_{\mathcal{M}^6} f(\mathbf{p}, \mathbf{x}, t) d^3p d^3x \tag{2.4}$$

gives the total number of particles. Note that the integral is clearly Lorentz invariant. The invariance of the distribution function itself is not obvious from such a definition and it has to be demonstrated explicitly, see Ref. [309]. From definition (2.4), one observes that $f(x,p)d^3pd^3x$ is an average number of particles having momenta in the range $(\mathbf{p}, \mathbf{p}+d^3p)$ and coordinates in the range $(\mathbf{x}, \mathbf{x}+d^3x)$ at the moment t, and the integral (2.4) is taken in the whole *phase space* $\mathcal{M}^6$. While the function $f(\mathbf{p}, \mathbf{x}, t)$ normalized to unity can be interpreted as a density of probability, this is not so for the function $\mathcal{F}(x,p)$ defined in Eq. (2.3); for discussion, see Ref. [156].

Note that, despite symmetrical form of $f(x,p)$, there is a conceptual difference between x and p. In particular, the integral

$$n(\mathbf{x}, t) = \int_{-\infty}^{+\infty} f d^3p \tag{2.5}$$

is assumed to be finite, leading to certain restrictions on $f(p)$. In particular, when the distribution function is isotropic in momentum space, $p^2 f(|p|)$ should decrease with increasing momentum for $|p| \gg 1$ fast enough, at least faster than $1/|p|$; it also should not increase with decreasing momentum for $|p| \ll 1$ faster than $1/|p|$.

For a function $a(x,p)$ depending on a microscopic state defined on the *phase space* $\mathcal{M}^6$, it is useful to introduce a *macroscopic quantity* $A(x)$ as

$$A(x) \equiv \int_{-\infty}^{+\infty} A(x,p) d^3p. \tag{2.6}$$

Particle density (2.5) is the simplest example. By definition, the *macroscopic quantity* does not depend on momentum but only on coordinates and time. In this way, one may construct moments of distribution function. It comes out that the first two moments of distribution function prove useful *macroscopic quantities* and help establish relation between kinetic theory and hydrodynamics, see Ref. [309].

Recalling Eq. (2.3), one can define an invariant quantity, instead of Eq. (2.5), as

$$j^{\mu}(\mathbf{x},t) \equiv c \int p^{\mu} f \frac{d^3p}{p^0} = c \int \mathcal{F}(x,p) p^{\mu} d^4p, \tag{2.7}$$

where both f and d^3p/p^0 are scalars. This *first moment* of the distribution function is the particle *four-flux*. Its spatial part represents usual three-vector flux

$$\mathbf{j}(\mathbf{x},t) = c \int \mathbf{v} f d^3p, \tag{2.8}$$

where $\mathbf{v} = c\mathbf{p}/p^0$ is the velocity vector of a relativistic particle with momentum $\mathbf{p}$, while its velocity four-vector is $u^{\mu} = dx^{\mu}/d\tau$.

Analogously, the *second moment* can be constructed as

$$T^{\mu\nu}(\mathbf{x},t) \equiv c \int p^{\mu} p^{\nu} f \frac{d^3p}{p^0}, \tag{2.9}$$

and so on. The quantity $T^{\mu\nu}$ is a symmetric tensor by construction. It represents an energy–momentum tensor of the system of particles.

2.2 Formulation of kinetic equation

First, assume for simplicity that particles do not interact. One can introduce a scalar quantity

$$\Delta J = \frac{1}{c} \int_{\Delta^3\sigma} d^3\sigma_{\mu} j^{\mu} = \int_{\Delta^3\sigma} d^3\sigma_{\mu} \int \frac{d^3p}{p^0} p^{\mu} f, \tag{2.10}$$

where the time-like four-vector $d^3\sigma_{\mu}$ is an oriented three-surface element of a plane *space-like surface* σ, the quantity $\Delta^3\sigma$ is a small element, and the last equality follows from Eq. (2.7). In the *Lorentz reference frame* where $d^3\sigma_{\mu}$ is purely time-like, it has components $(d^3x, 0, 0, 0)$. In this frame,

$$\Delta J = \int_{\Delta^3\sigma} \int f(x,p)\, d^3p d^3x, \tag{2.11}$$

which is just an average number of *world lines* crossing the segment $\Delta^3\sigma$. Considering those world lines, which have momenta in the range $\Delta^3 p$ around $\mathbf{p}$, one can get

$$\Delta J = \int_{\Delta^3\sigma}\int_{\Delta^3 p} f(x,p)\, d^3p d^3x. \tag{2.12}$$

Accepting this interpretation, consider *world lines* given by Eq. (2.10) which later cross another segment $\Delta^3\hat{\sigma}$. Since there are no collisions, it is possible to write

$$\int_{\Delta^3\hat{\sigma}} d^3\sigma_\mu \int_{\Delta^3 p} \frac{d^3p}{p^0} p^\mu f - \int_{\Delta^3\sigma} d^3\sigma_\mu \int_{\Delta^3 p} \frac{d^3p}{p^0} p^\mu f = 0 \tag{2.13}$$

or, in different form,

$$\int_{\Delta^3 x} d^3\sigma_\mu \int_{\Delta^3 p} \frac{d^3p}{p^0} p^\mu f = 0, \tag{2.14}$$

where $\Delta^3 x$ is the surface of *Minkowski space* element $\Delta^4 x$. Applying *Gauss theorem*, one gets

$$\int_{\Delta^4 x} d^4x \int_{\Delta^3 p} \frac{d^3p}{p^0} p^\mu \partial_\mu f = 0, \tag{2.15}$$

where $\partial_\mu = \left(c^{-1}\partial/\partial t, \nabla\right)$, $\Delta^3 x$ and $\Delta^3 p$ are some arbitrary hypersurfaces in the *phase space*.

The master equation represents time evolution of the distribution function due to microscopic interactions in the system. In the absence of any interactions between particles, it represents continuity of the four-vector $p^\mu f$ and it follows from Eq. (2.15) as

$$p^\mu \partial_\mu f = 0. \tag{2.16}$$

Written in the vector notation (and dividing by p^0/c),

$$\frac{df}{dt} \equiv \frac{\partial f}{\partial t} + \mathbf{v}\cdot\nabla f = 0. \tag{2.17}$$

In a general case, both collisions and external forces alter Eq. (2.16). Since such alteration is essentially local, the kinetic equation becomes

$$p^\mu \partial_\mu f + mF^\mu \frac{\partial f}{\partial p^\mu} = \mathrm{St}\,[f], \tag{2.18}$$

where F^{μ} represents an external four-force and St $[f]$ is the *collision integral*. This is the *relativistic transport equation*. One of the main goals of kinetic theory is to establish the form of the *collision integral*.

2.3 Collision integral for particle scattering

The simplest interaction between particles is scattering. Consider an elastic collision

$$1+2 \longrightarrow 1'+2', \tag{2.19}$$

where particles 1 and 2 have masses m_1 and m_2 and momenta p_1^{μ} and p_2^{μ} which changed after the collision to $p_1'^{\mu}$ and $p_2'^{\mu}$, respectively. Energy–momentum conservation gives

$$p_1^{\mu}+p_2^{\mu}=p_1'^{\mu}+p_2'^{\mu}. \tag{2.20}$$

The average number of such collisions is proportional to (1) the number of particles per unit volume with momenta p_1^{μ} in the range d^3p_1, (2) the number of particles per unit volume with momenta p_2^{μ} in the range d^3p_2, and (3) the intervals $d^3p_1'^{\mu}$, $d^3p_2'^{\mu}$ and d^4x. The proportionality coefficients, depending only on four-momenta before and after the collision, are represented as $W\left(p_1,p_2 \mid p_1',p_2'\right)/\left(p_1^0p_2^0p_1'^0p_2'^0\right)$. The quantity $W\left(p_1,p_2 \mid p_1',p_2'\right)$ is called the *transition rate* and it is a scalar. By this process, particles leave the phase volume d^3p_1 around p_1^{μ}. Collisions also bring particles back into this volume by the inverse process with the corresponding rate $W\left(p_1',p_2' \mid p_1,p_2\right)$.

Then the Boltzmann equation can be written as

$$\begin{aligned}\int_{\mathcal{V}}\int_{\mathcal{P}} p^{\mu}\partial_{\mu}f\frac{d^3p_1}{p_1^0}d^4x=\frac{1}{2}\int_{\mathcal{V}}\int_{\mathcal{P}}\int \frac{d^3p_1}{p_1^0}\frac{d^3p_1'}{p_1'^0}\frac{d^3p_2}{p_2^0}\frac{d^3p_2'}{p_2'^0}\\ \times\left[f\left(x,p_1'\right)f\left(x,p_2'\right)W\left(p_1',p_2' \mid p_1,p_2\right)\right.\\ \left.-f\left(x,p_1\right)f\left(x,p_2\right)W\left(p_1,p_2 \mid p_1',p_2'\right)\right]d^4x,\end{aligned} \tag{2.21}$$

where $\mathcal{V}$ and $\mathcal{P}$ are volumes in coordinate and momentum spaces, respectively, or in differential form

$$p^\mu \partial_\mu f = \frac{1}{2} \int \frac{d^3 p_1'}{p_1'^0} \frac{d^3 p_2}{p_2^0} \frac{d^3 p_2'}{p_2'^0} \times \left[f\left(x, p_1'\right) f\left(x, p_2'\right) W\left(p_1', p_2' \mid p_1, p_2\right) - f\left(x, p_1\right) f\left(x, p_2\right) W\left(p_1, p_2 \mid p_1', p_2'\right)\right] d^4 x. \quad (2.22)$$

The same equation in vector notation becomes

$$\frac{\partial f}{\partial t} + \mathbf{v} \cdot \nabla f = \frac{1}{2} \int d^3 p_1' d^3 p_2 d^3 p_2' \times \left[f\left(x, p_1'\right) f\left(x, p_2'\right) w_{p_1' p_2'; p_1 p_2} - f\left(x, p_1\right) f\left(x, p_2\right) w_{p_1 p_2; p_1' p_2'} \right], \quad (2.23)$$

where $w_{p_1 p_2; p_1' p_2'} = cW\left(p_1, p_2 \mid p_1', p_2'\right) / \left(p_1^0 p_2^0 p_1'^0 p_2'^0\right)$. If in this expression particle momenta are substituted by their velocities, this equation will coincide with the one derived first by Boltzmann [82]. Hence, the equation of the form (2.22) is called the Boltzmann equation. In what follows, also the more general equation of the form (2.18) is called the Boltzmann equation. Note that the factor 1/2 in front of *collision integral* is due to the fact that particles are indistinguishable.

2.4 Hydrodynamic equations

We start from the kinetic equation without collisions:

$$\frac{1}{c} \frac{\partial f}{\partial t} + v^\beta \frac{\partial f}{\partial x^\beta} = 0. \quad (2.24)$$

Let's write hydrodynamic equations from the kinetic equation (2.24). Let the coordinate system K' move with velocity $\mathbf{V}$ in the unmoving system K. The transformation of electron energy–momentum is (Mihalas)

$$\epsilon = \Gamma\left(\epsilon' + V\mathbf{N}\cdot\mathbf{p}'\right), \quad \mathbf{p} = \mathbf{p}' + \left[(\Gamma-1)\mathbf{N}\cdot\mathbf{p}' + \Gamma\frac{V}{c}\frac{\epsilon'}{c}\right]\mathbf{N}, \tag{2.25}$$

where

$$\Gamma = \frac{1}{\sqrt{1-(V/c)^2}}, \quad \mathbf{N} = \frac{\mathbf{V}}{V}.$$

The energy and the momentum relate with the mass and velocity in K

$$\epsilon = \gamma mc^2, \quad \mathbf{p} = \gamma m\mathbf{v}, \quad \gamma = \frac{1}{\sqrt{1-(v/c)^2}},$$

and the similar relations exist in the K' system.[1]

Let us select $\mathbf{e}_1$ along $\mathbf{N}$: $\mathbf{N} = (1,0,0)$. Then

$$dp_1 = dp_1' + (\Gamma-1)dp_1' + \Gamma\frac{V}{c}\frac{d\epsilon'}{c} = \Gamma\left(1 + \frac{c^2p_1'}{\epsilon'}N_1\right)$$

$$dp_1' = \frac{\epsilon}{\epsilon'}dp_1', \quad dp_{2,3} = dp_{2,3}',$$

$$d\mathbf{p} = \frac{\epsilon}{\epsilon'}d\mathbf{p}'.$$

Another relation is the invariant property of distribution function

$$f(\mathbf{p}) = f'(\mathbf{p}').$$

We adopt next assumption in the following. The distribution function for particles is locally equilibrium in the comoving coordinate

[1] To receive the inverse transformation, it is enough to change the sign of the vector $\mathbf{N}$:

$$\epsilon' = \Gamma(\epsilon - V\mathbf{N}\cdot\mathbf{p}), \tag{2.26}$$

$$\mathbf{p}' = \mathbf{p} - \left[-(\Gamma-1)\mathbf{N}\cdot\mathbf{p} + \Gamma\frac{V}{c}\frac{\epsilon}{c}\right]\mathbf{N}. \tag{2.27}$$

It is simply to receive above equations. From (2.25), after ϵ is multiplied by V/c^2 minus $\mathbf{p}$ multiplied by $\mathbf{N}$, one receives

$$\mathbf{N}\cdot\mathbf{p}' = -\Gamma(\epsilon - V\mathbf{N}\cdot\mathbf{p}).$$

Then from (2.25), one has Eqs. (2.26) and (2.27).

system K'. We want to describe particles by only three parameters: density, velocity, and energy. For example, for non-degenerate particles, the distribution function in comoving coordinates is

$$f'(\mathbf{p}') = \frac{g}{(2\pi\hbar)^3} \exp\left(-\frac{\epsilon' - \varphi}{kT}\right), \quad \epsilon' = \sqrt{m^2c^4 + (\mathbf{p}')^2c^2},$$

where g is a statistical weight.

Let's now define the hydrodynamic parameters of particles. The particle's and mass's densities are

$$n \equiv \int f' d\mathbf{p}', \quad \rho \equiv mn, \tag{2.28}$$

the pressure is

$$P \equiv c^2 \int f' \frac{(\mathbf{p}')^2}{3\epsilon'} d\mathbf{p}', \tag{2.29}$$

and the specific internal energy density is

$$\varepsilon \equiv \frac{1}{\rho} \int f' \epsilon' d\mathbf{p}', \tag{2.30}$$

where

$$d\mathbf{p}' = 4\pi(p')^2 dp', \quad \epsilon' d\epsilon'/c^2 = p' dp'. \tag{2.31}$$

Then using the relations (odd powers of $\mathbf{p}'$ give zero at the integration over $d\mathbf{p}'$ because $f(\mathbf{p}')$ is the even function),

$$m \int d\mathbf{p} f = m \int d\mathbf{p}' \frac{\Gamma(\epsilon' + V(\mathbf{N}\mathbf{p}'))}{\epsilon'} f'(\mathbf{p}') = \Gamma mn = \Gamma\rho,$$

$$\int d\mathbf{p}\mathbf{v} f = \int d\mathbf{p} \frac{\mathbf{p}}{\epsilon/c^2} f = \int d\mathbf{p}' \frac{\epsilon}{\epsilon'} \frac{\mathbf{p}}{\epsilon/c^2} f'$$

$$= \int d\mathbf{p}' \frac{\mathbf{p}' + \left((\Gamma - 1)(\mathbf{N}\mathbf{p}') + \Gamma\frac{V}{c}\frac{\epsilon'}{c}\right)\mathbf{N}}{\epsilon'/c^2} f' = \Gamma\mathbf{V}n,$$

$$\int d\mathbf{p}\mathbf{p}f = \int \frac{d\mathbf{p}'f'}{\epsilon'}\left(\mathbf{p}' + \left((\Gamma-1)(\mathbf{N}\mathbf{p}') + \Gamma\frac{V}{c}\frac{\epsilon'}{c}\right)\mathbf{N}\right)$$
$$\times\Gamma(\epsilon' + V(\mathbf{N}\mathbf{p}'))$$
$$= \int \frac{d\mathbf{p}'f'}{\epsilon'}\Gamma\left(\mathbf{p}'V(\mathbf{N}\mathbf{p}') + (\Gamma-1)(\mathbf{N}\mathbf{p}')^2 V\mathbf{N}\right.$$
$$\left. + \Gamma\frac{V}{c}\frac{\epsilon'^2}{c}\mathbf{N}\right)$$
$$= \Gamma^2\mathbf{V}\int d\mathbf{p}'f'\left(\frac{(\mathbf{p}')^2}{3\epsilon'} + \frac{\epsilon'}{c^2}\right) = \Gamma^2\rho\mathbf{V}\left(\frac{\varepsilon + \frac{P}{\rho}}{c^2}\right),$$

$$\frac{1}{c^2}\int d\mathbf{p}\epsilon f = \frac{1}{c^2}\int\frac{d\mathbf{p}'f'}{\epsilon'}\left(\Gamma(\epsilon' + V(\mathbf{N}\mathbf{p}'))\right)^2$$
$$= \frac{1}{c^2}\int\frac{d\mathbf{p}'f'}{\epsilon'}\Gamma^2\left(\epsilon'^2 + V^2(\mathbf{N}\mathbf{p}')^2\right)$$
$$= \frac{1}{c^2}\Gamma^2\int d\mathbf{p}'f'\left(\epsilon' + V^2\frac{(\mathbf{p}')^2}{3\epsilon'}\right)$$
$$= \Gamma^2\left(\frac{\rho\varepsilon}{c^2} + \left(\frac{\mathbf{V}}{c}\right)^2\frac{P}{c^2}\right),$$

$$\frac{1}{c^2}\int d\mathbf{p}p^\alpha v^\beta f = \int d\mathbf{p}\frac{p^\alpha p^\beta}{\epsilon}f$$
$$= \int d\mathbf{p}'\frac{f'}{\epsilon'}\left(p'^\alpha + \left[(\Gamma-1)\mathbf{N}\cdot\mathbf{p}' + \Gamma\frac{V}{c}\frac{\epsilon'}{c}\right]N^\alpha\right)$$
$$\times\left(p'^\beta + \left[(\Gamma-1)\mathbf{N}\cdot\mathbf{p}' + \Gamma\frac{V}{c}\frac{\epsilon'}{c}\right]N^\beta\right)$$
$$= \int d\mathbf{p}'\frac{f'}{\epsilon'}\left(p'^\alpha p'^\beta + \left[(\Gamma-1)\frac{2}{3}(\mathbf{p}')^2\delta^{\alpha\beta}\right.\right.$$
$$\left.\left. + (\Gamma-1)^2\frac{(\mathbf{p}')^2}{3} + \left(\Gamma\frac{V}{c}\frac{\epsilon'}{c}\right)^2\right]N^\alpha N^\beta\right)$$

$$= \int d\mathbf{p}' \frac{f'}{\epsilon'} \left(\frac{1}{3}(\mathbf{p}')^2 \delta^{\alpha\beta} + \left[(\Gamma - 1)\frac{2}{3}(\mathbf{p}')^2 \right.\right.$$

$$\left.\left. +(\Gamma - 1)^2 \frac{(\mathbf{p}')^2}{3} + \left(\Gamma \frac{V}{c}\frac{\epsilon'}{c} \right)^2 \right] N^\alpha N^\beta \right)$$

$$= \frac{P}{c^2}\delta^{\alpha\beta} + \left[2(\Gamma - 1)\frac{P}{c^2} + (\Gamma - 1)^2 \frac{P}{c^2} \right.$$

$$\left. + \Gamma^2 \frac{\rho\varepsilon}{c^2} + \left(\Gamma \frac{V}{c} \right)^2 \frac{\rho\varepsilon}{c^2} \right] N^\alpha N^\beta$$

$$= \frac{P}{c^2}\delta^{\alpha\beta} + \left[(\Gamma^2 - 1)\frac{P}{c^2} + \Gamma^2 \left(\frac{V}{c} \right)^2 \frac{\rho\varepsilon}{c^2} \right] N^\alpha N^\beta$$

$$= \frac{P}{c^2}\delta^{\alpha\beta} + \Gamma^2 \frac{P}{c^2}\frac{V^\alpha V^\beta}{c^2} + \Gamma^2 \frac{V^\alpha V^\beta}{c^2}\frac{\rho\varepsilon}{c^2}$$

$$= \frac{P}{c^2}\delta^{\alpha\beta} + \Gamma^2 \rho \frac{V^\alpha V^\beta}{c^2} \left(\frac{P}{\rho c^2} + \frac{\varepsilon}{c^2} \right),$$

$$m \int d\mathbf{p} \frac{\epsilon}{c^2} g^\beta \frac{\partial f}{\partial p^\beta} = -\frac{\mathbf{g}}{c^2}\Gamma V \rho,$$ [2]

$$\int d\mathbf{p}\epsilon \frac{\epsilon}{c^2} g^\beta \frac{\partial f}{\partial p^\beta} = -\mathbf{g} \int d\mathbf{p} 2\mathbf{p} f$$

$$= -2\mathbf{g}\Gamma^2 \rho \mathbf{V} \frac{\varepsilon + P/\rho}{c^2},$$

$$\int d\mathbf{p} p^\alpha \frac{\epsilon}{c^2}\frac{\partial f}{\partial p^\beta} = -\int d\mathbf{p} \frac{\partial (p^\alpha \epsilon / c^2)}{\partial p^\beta} f = -\int d\mathbf{p} f \left(\frac{\epsilon}{c^2}\delta^{\alpha\beta} + \frac{p^\alpha p^\beta}{\epsilon} \right)$$

$$= -\int d\mathbf{p}' f' \frac{\epsilon}{\epsilon'} \left(\frac{\epsilon}{c^2}\delta^{\alpha\beta} + \frac{p^\alpha p^\beta}{\epsilon} \right)$$

[2]

$$\epsilon^2 - \mathbf{p}^2 c^2 = m^2 c^4,$$

$$\epsilon \frac{\partial \epsilon}{\partial p^\beta} = p^\beta c^2.$$

$$
\begin{aligned}
&= -\int d\mathbf{p}' f' \frac{1}{\epsilon'} \left\{ \frac{[\Gamma(\epsilon' + V\mathbf{N}\cdot\mathbf{p}')]^2}{c^2} \delta^{\alpha\beta} \right. \\
&\qquad - \left(p'^{\alpha} + \left[(\Gamma - 1)\mathbf{N}\cdot\mathbf{p}' + \Gamma\frac{V}{c}\frac{\epsilon'}{c} \right] N^{\alpha} \right) \\
&\qquad \left. \times \left(p'^{\beta} + \left[(\Gamma - 1)\mathbf{N}\cdot\mathbf{p}' + \Gamma\frac{V}{c}\frac{\epsilon'}{c} \right] N^{\beta} \right) \right\} \\
&= -\int d\mathbf{p}' f' \frac{1}{\epsilon'} \left\{ \frac{(\Gamma\epsilon')^2 + (\Gamma V p')^2/3}{c^2} \delta^{\alpha\beta} \right. \\
&\qquad - \left(p'^{\alpha} p'^{\beta} + (\Gamma - 1) N^{\alpha} N^{\beta} ((p'^{\alpha})^2 + (p'^{\beta})^2) \right) \\
&\qquad \left. - \left[(\Gamma - 1)^2 \frac{(\mathbf{p}')^2}{3} + \left(\Gamma\frac{V}{c}\frac{\epsilon'}{c} \right)^2 \right] N^{\alpha} N^{\beta} \right\} \\
&= -\Gamma^2 \left(\frac{\rho\varepsilon}{c^2} + \left(\frac{V}{c} \right)^2 \frac{P}{c^2} \right) \delta^{\alpha\beta} \\
&\quad - \int d\mathbf{p}' f' \frac{1}{\epsilon'} N^{\alpha} N^{\beta} \left\{ \left(\frac{1}{3} (\mathbf{p}')^2 \delta^{\alpha\beta} + (\Gamma - 1) \frac{2}{3} (\mathbf{p}')^2 \right) \right. \\
&\qquad \left. - \left[(\Gamma - 1)^2 \frac{(\mathbf{p}')^2}{3} + \left(\Gamma\frac{V}{c}\frac{\epsilon'}{c} \right)^2 \right] \right\} \\
&= -\Gamma^2 \left(\frac{\rho\varepsilon}{c^2} + \left(\frac{V}{c} \right)^2 \frac{P}{c^2} \right) \delta^{\alpha\beta} \\
&\quad - \frac{P}{c^2} \left(\delta^{\alpha\beta} + ((\Gamma - 1)^2 + 2(\Gamma - 1)) N^{\alpha} N^{\beta} \right) \\
&\quad - \frac{\rho\varepsilon}{c^2} \Gamma^2 \frac{V^{\alpha} V^{\beta}}{c^2} \\
&= -\Gamma^2 \rho \left(\frac{P}{\rho c^2} + \frac{\varepsilon}{c^2} \right) \left(\delta^{\alpha\beta} + \frac{V^{\alpha} V^{\beta}}{c^2} \right),
\end{aligned}
$$

from Boltzmann equation (2.24), it is possible to obtain hydrodynamic equations — $m\int d\mathbf{p}\times$ Eq. (2.24):

$$\frac{\partial(\Gamma\rho)}{\partial t}+\frac{\partial}{\partial x^\alpha}(\Gamma\rho V^\alpha)=0,$$

$$\frac{\partial(\Gamma\rho)}{\partial t}+\frac{\partial}{\partial x^\alpha}(\Gamma\rho V^\alpha)=0, \tag{2.32}$$

$\int d\mathbf{pp}\times$ Eq. (2.24):

$$\frac{\partial}{\partial t}\left(\Gamma\rho\Gamma hV^\alpha\right)+\frac{\partial}{\partial x^\beta}\left(\Gamma\rho\Gamma hV^\alpha V^\beta+\delta^{\alpha\beta}P\right)=0,$$

$$\frac{\partial}{\partial t}\left(\Gamma\rho\Gamma hV^\alpha\right)+\frac{\partial}{\partial x^\beta}\left(\Gamma\rho\Gamma hV^\alpha V^\beta\right)+\frac{\partial}{\partial x^\beta}\left(\delta^{\alpha\beta}P\right)=0, \tag{2.33}$$

$\int d\mathbf{p}\epsilon\times$ Eq. (2.24):

$$\frac{\partial}{\partial t}\left(\Gamma\rho\left(\Gamma c^2h-\frac{P}{\Gamma\rho}\right)\right)+\frac{\partial}{\partial x^\alpha}\left(\Gamma\rho V^\alpha\Gamma c^2h\right)=0,$$

$$\frac{\partial}{\partial t}\left(\Gamma\rho\left(\Gamma c^2h-\frac{P}{\Gamma\rho}\right)\right)+\frac{\partial}{\partial x^\alpha}\left(\Gamma\rho V^\alpha\Gamma c^2h\right)=0, \tag{2.34}$$

$$\frac{\partial}{\partial t}\left(\Gamma\rho\left(\Gamma c^2h-\frac{P}{\Gamma\rho}\right)\right)+\frac{\partial}{\partial x^\alpha}\left(\Gamma\rho V^\alpha\Gamma c^2h\right)=0.$$

The specific dimensionless enthalpy is

$$h=\frac{1}{c^2}\left(\varepsilon+\frac{P}{\rho}\right).$$

Hydrodynamical equations are (2.32)–(2.34). For our applications in SN physics, the non-relativistic case is appropriate with $V\ll c$.

Chapter 3

Equation of State of the Matter at High Density

3.1 Equation of state of non-interacting particles for a different type of statistics

Statistical mechanics operates with a huge number of particles. The canonical ensemble gives the probability of the system X being in state x (equivalently, of the random variable X having value x) as

$$P(x) = \frac{\exp(-\beta E(x))}{Z(\beta)}. \tag{3.1}$$

Here, $E(x)$ is the energy of the configuration x. The parameter $\beta = (k_{\mathrm{B}}T)$ is the inverse temperature. The statistical analysis gives for the non-interacting particles (so cold an ideal gas) an average number of particles in the state with the energy ϵ_k

$$\bar{n}_k = k_{\mathrm{B}}T\frac{\partial}{\partial\mu}\ln\sum_{n_k=0}^{N}\left(e^{\frac{\mu-\epsilon_k}{k_{\mathrm{B}}T}}\right)^{n_k} \tag{3.2}$$

with the *chemical potential of particles* μ. In the case of $e^{\frac{\mu-\epsilon_k}{k_{\mathrm{B}}T}} < 1$, one has the geometric progression ($\mu \leq 0$), and

$$\sum_{n_k=0}^{\infty}\left(e^{\frac{\mu-\epsilon_k}{k_{\mathrm{B}}T}}\right)^{n_k} = \frac{1}{1-e^{\frac{\epsilon_k-\mu}{k_{\mathrm{B}}T}}} \tag{3.3}$$

and

$$\bar{n}_{=}\frac{1}{e^{\frac{\epsilon_k-\mu}{k_{\mathrm{B}}T}}-1}. \tag{3.4}$$

It is the so-called *Bose–Einstein statistic* case. In the case of Pauli blocking, one has only a limited set of the number $n_k = 0$ and 1, and *Fermi–Dirac statistic*

$$\bar{n}_{=} = k_{\mathrm{B}}T\frac{\partial}{\partial\mu}\ln\left(1+e^{\frac{\mu-\epsilon_k}{k_{\mathrm{B}}T}}\right) = \frac{1}{e^{\frac{\epsilon_k-\mu}{k_{\mathrm{B}}T}}+1}. \tag{3.5}$$

The distribution function in the phase space $(\mathbf{r}, \mathbf{p})$ for for Fermi–Dirac and Bose–Einstein statistics is

$$f(\mathbf{p}) = \frac{g}{(2\pi\hbar)^3}\frac{1}{\exp((\epsilon-\mu)/(k_{\mathrm{B}}T))\pm 1}, \tag{3.6}$$

taking into account limited number of quantum states and corresponding statistical weights g. It is interesting to mention that the same result can be obtained in direct numerical simulations of the evolution of arbitrary initial uniform distribution functions for pairs (Fermi–Dirac) and photons (Bose–Einstein) to the equilibrium state. Usually, the system interacts fast in binary interactions, the so-called kinetic equilibrium with non-zero chemical potentials, and equal temperatures are established. In the long time, the three particle interactions change the number of particles corresponding to the thermal equilibrium with zero chemical potentials and the temperature corresponds to the conserve energy density.

Using the definitions of parameters of the equation of state (2.28)–(2.30), we obtain explicit expressions for concentration

$$n = \frac{g}{(2\pi\hbar)^3}\int\frac{1}{\exp\left[\frac{\epsilon-\mu}{k_{\mathrm{B}}T}\right]\pm 1}\frac{4\pi\sqrt{\epsilon^2-m^2c^4}\epsilon d\epsilon}{c^3}, \tag{3.7}$$

for pressure

$$P = \frac{g}{(2\pi\hbar)^3}\int\frac{1}{\exp\left[\frac{\epsilon-\mu}{k_{\mathrm{B}}T}\right]\pm 1}\frac{4\pi(\epsilon^2-m^2c^4)^{3/2}d\epsilon}{3c^3}, \tag{3.8}$$

and for internal energy density

$$\rho\varepsilon = \frac{g}{(2\pi\hbar)^3}\int \frac{1}{\exp\left[\frac{\epsilon-\mu}{k_{\mathrm{B}}T}\right]\pm 1}\frac{4\pi\sqrt{\epsilon^2-m^2c^4}\epsilon^2 d\epsilon}{c^3}. \tag{3.9}$$

3.1.1 *Non-relativistic and ultra-relativistic particles*

For the *non-relativistic case*, we have

$$p \approx mv, \quad \epsilon \approx mc^2 + \tfrac{p^2}{2m}, \quad \sqrt{\epsilon^2 - m^2c^4} = cp,$$
$$d(\epsilon - mc^2) \approx \tfrac{pdp}{m}, \tag{3.10}$$

for internal energy density without taking into account the specific internal energy c^2,

$$\rho\varepsilon - nmc^2 = \frac{g}{(2\pi\hbar)^3}\int \frac{1}{\exp\left[\frac{\epsilon-\mu}{k_{\mathrm{B}}T}\right]\pm 1}\frac{4\pi\sqrt{\epsilon^2-m^2c^4}\epsilon(\epsilon-c^2)d\epsilon}{c^3}, \tag{3.11}$$

and for pressure

$$P = \frac{g}{(2\pi\hbar)^3}\int \frac{1}{\exp\left[\frac{\epsilon-\mu}{k_{\mathrm{B}}T}\right]\pm 1}\frac{4\pi\sqrt{\epsilon^2-m^2c^4}(\epsilon-mc^2)}{c^3}\frac{\epsilon+mc^2}{3}d\epsilon. \tag{3.12}$$

Taking into account the relations $pc \ll mc^2$ and $(\epsilon + mc^2) \approx 2\epsilon$ one has for hydrodynamical equations (for independent variables, density, momentum, and energy density), the equation of state looks like as an ideal gas law with the adiabatic index $\gamma = 5/3$ for any degree of a degeneracy

$$P = (\gamma - 1)\rho(\epsilon - c^2), \quad \gamma \approx \frac{5}{3}. \tag{3.13}$$

Of course, the relation between density, temperature, and internal energy will be an ideal gas law EOS only non-degenerate case, then $\exp(\epsilon-\mu)/(k_{\mathrm{B}}T) \gg 1$. At the gravitational collapse, nucleon reaches nuclear densities and becomes degenerate, but due to a huge mass 940 MeV, they are non-relativistic.

For *ultra-relativistic case* we have in ultra-relativistic case, we have

$$\epsilon \gg mc^2, \quad \epsilon = cp, \tag{3.14}$$

and for internal energy density,

$$\rho\varepsilon = \frac{g}{(2\pi\hbar)^3} \int \frac{1}{\exp\left[\frac{\epsilon-\mu}{k_{\rm B}T}\right] \pm 1} \frac{4\pi\epsilon^3 d\epsilon}{c^3}, \tag{3.15}$$

and for pressure,

$$P = \frac{g}{(2\pi\hbar)^3} \int \frac{1}{\exp\left[\frac{\epsilon-\mu}{k_{\rm B}T}\right] \pm 1} \frac{4\pi\epsilon^3 d\epsilon}{3c^3}. \tag{3.16}$$

For hydrodynamics, the equation of state looks like as an ideal gas law with the adiabatic index $\gamma = 4/3$ for any degree of a degeneracy

$$P = (\gamma - 1)\rho(\epsilon - c^2), \quad \gamma \approx \frac{4}{3}. \tag{3.17}$$

Electrons and positrons at the gravitational collapse and in the compact stars like white dwarfs are degenerate and already relativistic due to relatively small electron mass 0.5 MeV. They have chemical potential with opposite signs $\mu_{e^+} = -\mu_{e^-}$, $\mu_e \equiv \mu_{e^-}$. For ultra-relativistic ($\mu_e \gg m_e c^2$) and degenerate ($k_{\rm B}T \ll \rho\epsilon/m_p$), the electron–positron pair EOS is

$$(n_{e^-} - n_{e^+}) = \frac{1}{2\pi^2}\left(\frac{kT}{\hbar c}\right)^3 (\mu_e^3 + \pi^2\mu_e), \tag{3.18}$$

$$P_e = \left(\frac{4}{3} - 1\right)\rho\epsilon_e = \frac{1}{12\pi^2(\hbar c)^3}$$

$$\left(\mu_e^4 + 2\pi^2\mu_e^2(k_{\rm B}T)^2 + \frac{7}{15}\pi^4(k_{\rm B}T)^4\right). \tag{3.19}$$

3.2 Nuclear statistical equilibrium and other models of EOS

The nuclear statistical equilibrium model of EOS (NSE) is appropriate in the initial state of collapse and then degenerate relativistic electrons play a main role in EOS. Moreover, such EOS is suitable to study of conditions of large-scale convection due to neutronization in the center.

The equation of state (EOS) of matter represents coupled equations for pressure ($P_{\rm m}$) and internal energy ($\varepsilon_{\rm m}$) as functions of density, temperature, and in our case of a week interaction also the number of electrons ($Y_{\rm e}$):

$$\begin{aligned} P_{\rm m} &= P_{\rm m}(\rho, \varepsilon_{\rm m}, Y_{\rm e}), \\ \varepsilon_{\rm m} &= \varepsilon_{\rm m}(\rho, T, Y_{\rm e}). \end{aligned} \tag{3.20}$$

Matter includes nuclei with free nucleons in statistical equilibrium, pairs, and equilibrium radiation. The statistical equilibrium nuclei (mass numbers A_i and charges Z_i) with free nucleons,

$$(A_i, Z_i) \rightleftarrows (A_i - Z_i)n + Z_i p, \tag{3.21}$$

give the relation of chemical potentials

$$\mu_i = (A_i - Z_i)\mu_n + Z_i \mu_p, \tag{3.22}$$

where the chemical potential i-nuclei can be defined in non-relativistic case (taking into account degeneracy of Fermi particles) from relation

$$\frac{X_i \rho}{A_i m_p} = 4\pi\omega_i \left(\frac{m_i c}{2\pi\hbar}\right)^3 \frac{1}{2} \left(\frac{2k_{\rm B}T}{m_i c^2}\right)^{3/2} F_{1/3}\left(\frac{\mu_i - m_i c^2}{k_{\rm B}T}\right) \tag{3.23}$$

with number densities of nuclei X_i, Fermi–Dirac function

$$F_\alpha(x) \equiv \int_0^\infty \frac{\xi^\alpha d\xi}{1 + \exp(\xi - x)} \tag{3.24}$$

of order $\alpha = 1/3$, and statistical weight ω_i. The sum of number densities equals to one

$$\sum_i X_i = 1, \tag{3.25}$$

while the electroneutrality gives the relation

$$\sum_i \frac{X_i Z_i}{A_i} = Y_e. \tag{3.26}$$

Pressure and specific internal energy of nucleons are described by equations

$$P_i = \frac{4\pi\omega_i m_i c^3}{3}\left(\frac{2k_\mathrm{B}T}{m_i c^2}\right)^3 \frac{1}{2}F_{3/2}\left(\frac{\mu_i - m_i c^2}{k_\mathrm{B}T}\right), \tag{3.27}$$

$$\varepsilon_i = \frac{m_i c^2 X_i}{A_i m_p} + \frac{4\pi\omega_i m_i c^3}{\rho}\frac{1}{4}\left(\frac{2k_\mathrm{B}T}{m_i c^2}\right)^3 F_{3/2}\left(\frac{\mu_i - m_i c^2}{k_\mathrm{B}T}\right). \tag{3.28}$$

For nuclei, it is not necessary to take into account degeneracy, as it can be described by ideal gas law EOS. Radiation is the black body.

For electrons, one can use EOS of ultra-relativistic pairs ($\mu_\mathrm{e} \gg m_\mathrm{e}c^2$):

$$\begin{aligned}\frac{Y_\mathrm{e}\rho}{m_\mathrm{p}} &= \frac{1}{2\pi^2}\left(\frac{kT}{\hbar c}\right)^3(\mu_\mathrm{e}^3 + \pi^2\mu_\mathrm{e}),\\ P_\mathrm{e} &= \left(\frac{4}{3} - 1\right)\rho\epsilon_\mathrm{e}\\ &= \frac{1}{12\pi^2(\hbar c)^3}\left(\mu_\mathrm{e}^4 + 2\pi^2\mu_\mathrm{e}^2(k_\mathrm{B}T)^2 + \frac{7}{15}\pi^4(k_\mathrm{B}T)^4\right). \end{aligned} \tag{3.29}$$

It is interesting to note that an ideal gas relation between pressure and internal energy for ultra-relativistic Fermi particles is true for any degree of degeneracy

$$P_\mathrm{e} = (\Gamma - 1)\,\rho\epsilon_\mathrm{e} \tag{3.30}$$

with adiabatic index $\Gamma = 4/3$. In non-relativistic case at any degree of degeneracy, the adiabatic index is $5/3$. Without reactions and neutrino transport, the ideal conservative hydrodynamic is a suitable approach [17].

Due to high matter density of a collapsing core, photons, electrons, positrons, nucleons, and nuclei are described by equilibrium distribution functions. Photons are in thermal equilibrium with a temperature and zero chemical potential

$$P_\gamma = \frac{\pi^2(k_\mathrm{B}T)^4}{3 \cdot 15(c\hbar)^3} = (4/3-1)\rho\epsilon_\gamma \tag{3.31}$$

and entropy per baryon $s = 4\frac{\pi^2(k_\mathrm{B}T)^4}{3 \cdot 15(c\hbar)^3} T^3/3m_p/(k_\mathrm{B}\rho)$. Due to two-particle and three-particle electromagnetic interactions, total energy density (or equilibrium temperature) defines not only the photons energy density but also the concentration of photons. But electrons, positrons, nucleons, and nuclei take part in week interactions with neutrino conserving the so-called lepton charge. For this reason, pairs has non-zero chemical potential. Electrons, positrons, protons, and nuclei obey only electroneutrality condition.

The important characteristic of EOS is entropy of the matter per nucleon, for pairs

$$s_\mathrm{e} = \frac{m_\mathrm{p}}{k_\mathrm{B}\rho} \frac{\partial P_\mathrm{e}(\rho, T, Y_\mathrm{e})}{\partial T}, \tag{3.32}$$

where $P_{\mathrm{e},T} = \frac{1}{12\pi^2(\hbar c)^3}(4\mu_\mathrm{e}^3\mu_{\mathrm{e}T} + 2\pi^2 2\mu_\mathrm{e}\mu_{\mathrm{e},T}(k_\mathrm{B}T)^2 + 2\pi^2\mu_\mathrm{e}^2 2(k_\mathrm{B}T)k_\mathrm{B} + \frac{7}{15}\pi^4 4(k_\mathrm{B}T)^3 k_\mathrm{B})$, $\mu_{\mathrm{e},T} = -\frac{\pi^2\mu_\mathrm{e}2k_\mathrm{B}*Tk_\mathrm{B}}{(3\mu_\mathrm{e}^2+\pi^2(k_\mathrm{B}T)^2)} + \mu_\mathrm{e}\left(\frac{k_\mathrm{B}T}{k_\mathrm{B}T+0.25m_\mathrm{e}c^2)}\right)^2 \frac{0.25m_\mathrm{e}c^2}{c_\mathrm{B}T^2}$. Entropy of nucleon type i with concentration x_i is $s_i = (P_i + \rho\epsilon_i - x_i * \rho/(A_i * m_\mathrm{p})(m_i * c^2 + \log(x_i\rho/(\omega_i A_i m_\mathrm{p})(2\pi\hbar^2/(m_i k_\mathrm{B}T))^{1.5})k_\mathrm{B}T))/T * m_\mathrm{p}/(k_\mathrm{B}\rho)$.

In the conservative scheme, the independent variables are $(\rho, \varepsilon_\mathrm{m}, Y_\mathrm{e})$, so one needs to resolve matter temperature, T, from a nonlinear EOS (3.20) by a Newton iteration taking into account electroneutrality condition (3.26).

The number of electrons Y_e is determined from the kinetics of neutronization due to weak interactions

$$\frac{\partial Y_\mathrm{e}}{\partial t} + \mathbf{v}\nabla Y_\mathrm{e} = \dot{Y}_\mathrm{e}(\rho, T_\mathrm{m}, \varepsilon_\nu). \tag{3.33}$$

Strictly speaking, in EOS, only pair plasma is ultra-relativistic. Nucleons are not relativistic but become degenerate at high densities. Nuclei are not relativistic and they are not degenerate.

To carry out the interaction of neutrinos with the shell, one should significantly complicate calculations by the presented implicit scheme. Moreover, the replacement of nuclear statistical equilibrium by more accurate EOS at high density is important for the accurate calculation of the forming of a neutron star and for the test of the neutrino-driven mechanism of SN. For example, the authors [192] considered the two-nucleon nuclear interaction in EOS. For core collapse simulations, Lattimer & Swesty EOS is widely used [206,207]. In simplified approach, it is possible to add the term corresponding to stiff EOS of baryons at high density

$$\begin{aligned} P_{\mathrm{BHD}} &= 364\ \mathrm{MeV}/10^{-39}(10^{-39}\rho/m_{\mathrm{p}})^{2.54}, \\ \epsilon_{\mathrm{BHD}} &= P/((2.54-1)\rho), \end{aligned} \tag{3.34}$$

and zero entropy.

Chapter 4

Transport with Reactions

4.1 The neutrino transport in SN and weak interactions

Probably, the first self-consistent model and computations of the gravitational collapse in 1D were performed in Ref. [109]. The authors even demonstrated the deposition of the necessary amount of the energy in the pre-SN envelope, but they overestimated the neutrino transparency. In the Lagrangian variables $m = \int_0^r (r')^2 \rho(r', t) dr'$ and t, the hydrodynamics equations can be written in the following form, with accuracy to within terms of order v/c in a 1D Newtonian-gravity approximation [91]:

$$\frac{\partial r}{\partial t} = v, \tag{4.1}$$

$$\frac{\partial v}{\partial t} + r^2 \frac{\partial}{\partial m}\left(P - \zeta \frac{\partial (r^2 v)}{\partial V}\right) = -\frac{4\pi G m}{r^2} + \frac{1}{\rho c} \sum_{\nu q} \int d\mu d\epsilon_\nu \mu (\chi_\nu^q E_\nu - \eta_\nu^q), \tag{4.2}$$

$$\frac{\partial \epsilon}{\partial t} + \left(P - \zeta \frac{\partial (r^2 v)}{\partial V}\right) \frac{\partial (r^2 v)}{\partial m} = \frac{1}{\rho} \sum_{\nu} \int d\mu d\epsilon_\nu (\chi_\nu^q E_\nu - \eta_\nu^q), \tag{4.3}$$

where ϵ is the specific internal energy of the matter, ζ is the coefficient of artificial viscosity, and $V = r^3/3$. The following transport equation is valid for the intensity of neutrinos of sort ν $I_\nu(r, \mu, \epsilon_\nu, t)$ in the direction whose angle with the radius vector corresponds to the cosine

μ (in process q):

$$\frac{1}{c}\frac{\partial E_\nu}{\partial t} + \mu\frac{\partial r^2 E_\nu}{\partial V} + \frac{1}{r}\frac{\partial}{\partial \mu}\left\{(1-\mu^2)\left[1+\left(\frac{3v}{c}-\frac{r}{c}\frac{\partial r^2 v}{\partial V}\right)\mu\right]E_\nu\right\}$$
$$+\frac{1}{r}\left[\mu^2\left(\frac{3v}{c}-\frac{r}{c}\frac{\partial r^2 v}{\partial V}\right)-\frac{v}{c}\right]\frac{\partial \epsilon_\nu E_\nu}{\partial \epsilon_\nu}$$
$$+\frac{1}{r}\left\{\frac{v}{c}+\frac{r}{c}\frac{\partial r^2 v}{\partial V}-\mu^2\left(\frac{3v}{c}-\frac{r}{c}\frac{\partial r^2 v}{\partial V}\right)\right\}E_\nu = -\chi_\nu^q E_\nu + \eta_\nu^q. \tag{4.4}$$

Instead of the intensity I_ν, we use the spectral phase density of the energy $E_\nu(r,\epsilon,\mu,t) = 2\pi\epsilon_\nu^3 f_\nu/c^3 = 2\pi I_\nu/c$, where $f_\nu(\mathbf{r},\mathbf{p},t)$ and I_ν are distribution function and intensity for neutrinos of sort ν. For our spherically symmetrical case, an element of the phase volume is $dV d\epsilon_\nu d\mu$, and the energy of particles of type ν in such an element is $\int dV d\epsilon_\nu d\mu E_\nu$. The emission and absorption coefficients η_ν^q, χ_ν^q for neutrinos of type ν are calculated for each process q. Moreover, we must solve the kinetics equation for the difference in the number of electrons and positrons per nucleon Y_e:

$$\frac{\partial Y_{\mathrm{e}}}{\partial t} = \dot{Y}_{\mathrm{e}}(\rho, T, Y_{\mathrm{e}}, I_\nu). \tag{4.5}$$

To solve the system (4.1)–(4.5), we must have an expression for the emission and absorption coefficients on the right-hand sides of (4.5) and must specify the equation of state, $P = P(\rho, T, Y_e)$, described in Section 3.2.

4.2 Reaction rates

Necessary reactions of week interactions include scattering of neutrinos on electrons, absorption of neutrinos by neutrons, creation of neutrinos, absorption and emission of neutrinos by nuclei, corresponding reactions for antineutrino, neutrino–anti-neutrino annihilations if pairs, and pairs annihilations in neutrinos.

All considered reactions with neutrino are two-particle interactions. Weak interactions do not change the number of particles, such interactions conserve the so-called lepton charge. For this reason in comparison with the photons with zero chemical potential in thermal due to 2p- and 3p-interactions, we have non-zero chemical potential of electrons and positrons and equation for $Y_{\rm e}$.

Let us recall the definitions of the matrix elements of processes and the associated probabilities, scattering amplitudes, and differential cross-sections [43]. We will work in CGS units and retain the dimensional quantities $\hbar$ and c in formulas [59]. If there are a product particles and b reactant particles, the differential probability for the processes per unit time can be written as

$$dw = c(2\pi\hbar)^4\delta^{(4)}(\mathfrak{p}_{\rm f} - \mathfrak{p}_i)|M_{fi}|^2 V \left[\prod_b \frac{\hbar c}{2\epsilon_b V}\right] \left[\prod_a \frac{V d\mathbf{p}'_a}{(2\pi\hbar)^3} \frac{\hbar c}{2\epsilon'_a V}\right],$$

where $\mathbf{p}'_a$ and ϵ'_a are the momentum and energy of the product particles, ϵ_b is the energy of the particles before their interaction, M_{fi} is the matrix element for the interaction, $\delta^{(4)}$ denotes the conservation of energy–momentum, and V is the volume. The matrix element is convenient to use, first, because this quantity is dimensionless. This means that it is simpler to change from the often-used system of units in which $\hbar, c = 1$ to CGS units. Second, the matrix element contains the scalar products of the energy–momentum vectors, i.e., it is invariant. The scattering amplitude is related to the (dimensionless) *matrix element* as

$$T_{fi} = \sqrt{\left[\prod_b \frac{\hbar c}{2\epsilon_b V}\right] \left[\prod_a \frac{\hbar c}{2\epsilon'_a V}\right]} M_{fi}.$$

The probability for a transition per unit time is

$$w_{fi} = c(2\pi\hbar)^4\delta^{(4)}(\mathfrak{p}_f - \mathfrak{p}_i)|T_{fi}|^2 V.$$

The scattering amplitude is related to the scattering matrix as S_{fi}:

$$S_{fi} = \delta_{fi} + i(2\pi\hbar)^4\delta^{(4)}(\mathfrak{p}_f - \mathfrak{p}_i)T_{fi},$$

where δ_{fi} is the unit matrix.

In the important case of double collisions (two reactant particles and two product particles), the differential cross-section of the process $d\sigma$ is introduced:

$$dw = jd\sigma, \ j = \frac{cI}{\epsilon_1 \epsilon_2 V}, \quad I = c\sqrt{(\mathfrak{p}_1 \mathfrak{p}_2)^2 - (m_1 m_2 c^2)^2}.$$

It is not difficult to recalculate the matrix elements in terms of the differential cross-section, which is usually calculated in the center-of-mass system.

All weak interactions in detail on a computational grid are considered in Appendix B.

4.3 Thermonuclear burning as the engine of Type I SN

The thermonuclear explosion of degenerate carbon–oxygen core is Type I SN [37,83,100]. Main-sequence stars with mass $\lesssim 10\, M_{\odot}$ experience such an explosion, whereas the formation of an iron core and its collapse occurs in massive stars with mass $\gtrsim 10\, M_{\odot}$ [55]. Also evolution in the close binary with a process of mass exchange between the companions dramatically changes the mass ratio of carbon–oxygen stellar cores [293] and can reduce the masses of stars in binary. At least, half of a star is a binary system. If one star with mass $\sim 8\, M_{\odot}$ evaluates in a result of thermonuclear burning to a compact carbon–oxygen white dwarf, it can attract a matter from a companion. As a result of a mass attraction, a carbon–oxygen white dwarf becomes more compact, increases its temperature, and explores.

Hydrodynamic calculations of the thermonuclear explosion of the degenerate carbon–oxygen cores give us complete disruption of the whole star, with no gravitationally bound remnant [37,83,100, 185]. However, such a simplified approach with the assumption of the detonation regime of burning left almost no opportunity for a collapse to occur. This result contradicts to the correlation between neutron stars and SNe remnant [325]. Within the framework of the deflagration regime of carbon thermonuclear burning, it is possible to form a neuron star [185].

The central problem in a theory of the thermonuclear burning in a star is a possibility of the detonation in a carbon–oxygen matter with a degenerate electrons. In the case of an ideal gas law equation of state, one has strong correlation between the jump of pressure on a shock wave and the increasing of temperature which is important for the reaction rate. For this reason, the formation of a detonation wave is possible. In the case of degenerate electrons, the jump of a pressure correlates with the increase in density, and because the pressure mainly depends from density, the temperature increases slowly. The detonation can decay on deflagration [172].

The system of gas-dynamic conservation laws in orthogonal coordinates includes mass conservation equation

$$\frac{\partial \rho}{\partial t} + \mathrm{div}\rho\mathbf{v} = 0, \tag{4.6}$$

momentum conservation equation

$$\frac{\partial \rho\mathbf{v}}{\partial t} + \mathrm{Div}\Pi = -\mathbf{g}\rho, \tag{4.7}$$

with the tensor $\Pi_{ij} = \rho v_i v_j + P\delta_{ij}$, energy equation

$$\frac{\partial \rho E}{\partial t} + \mathrm{div}(\rho E + P)\mathbf{v} = \mathbf{g}\rho\mathbf{v} + \rho Q, \tag{4.8}$$

$$\frac{\partial}{\partial t}\rho\left(\frac{u^2}{2} + E\right) + \frac{1}{r^2}\frac{\partial}{\partial r}r^2 u\left(\rho\frac{u^2}{2} + \rho E + P\right) = \frac{GM}{r^2}\rho u + \rho Q, \tag{4.9}$$

with the energy density $E = \epsilon + \mathbf{v}^2/2$, and kinetic equation for the abundance of the element with atomic number i, $Y_i \equiv X_i/A_i$ with X_i and A_i the mass fraction and the molar mass of nuclide i, respectively,

$$\frac{\partial \rho Y_i}{\partial t} + \frac{1}{r^2}\frac{\partial r^2 \rho u Y_i}{\partial r} = \rho W_i, \tag{4.10}$$

where W_i is the rate of formation (exhaustion) of the i element in all reactions.

The change of *nuclear abundances* with time due to reactions is a system of ordinary differential equations [204]:

$$\dot{Y}_i = \rho\sum_{k,l\neq i} Y_k Y_l R_{kl,i} - \rho\sum_{l} Y_i Y_l R_{il,m} + \sum_{n\neq i} Y_n \lambda_{n,i} - Y_i\lambda_{i,k}. \tag{4.11}$$

All of the subscripts denote the sort of the nuclide: $R_{kl,i}$ — the rate of the reaction $k + l \to i + \cdots$ and $\lambda_{n,i}$ — the probability of photodisintegration $n \to i + {}^4$He per second. In calculations, the reaction rates usually use the next interpolation for temperature $T_9 \equiv T/10^9$ K with coefficients a_i for experimental data and theoretical modes[1]:

$$\propto \exp\left(a_0 + \frac{a_1}{T_9} + \frac{a_2}{T_9^{1/3}} + a_3 T_9^{1/3} + a_4 T_9 + a_5 T_9^{5/3} + a_6 \log T_9\right). \tag{4.12}$$

The energy release Q follows from Einstein's relation $E = \Delta M c^2$, between energy E, and the masses balance of input and outgoing particles, ΔM.

[1]See https://reaclib.jinaweb.org/.

Part 2
Numerical Methods

Chapter 5

The Basics of Computational Physics

This chapter provides a brief informal introduction in computational physics. While analytic results contain all information, often it is impossible to obtain analytic solutions of complex differential equation, especially nonlinear ones. Then new results can be revealed from numerical simulations.

The numerical method is the important ingredient for the computational simulations. There is no universal set of numerical receipts applicable for the whole range of physical problems. Nevertheless, some numerical algorithms now make part of such computational programs as Mathematica[1] or MATLAB[2] [317]. It has to be emphasized that basic ideas in computational physics come from the theoretical and mathematical physics.

This chapter describes the standard types of equations of classical mathematical physics along with methods of their solutions. The focus is on finite difference methods. The system of ordinary differential equations and problems of linear algebra are also considered. In addition, the order of accuracy and the stability of the scheme, providing the convergence of the numerical solution to the actual solution of the partial differential equation, are discussed.

[1] https://www.wolfram.com/mathematica/.

[2] https://www.mathworks.com/products/matlab/.

5.1 Finite differences and computational grids

The computational technology cannot represent the smooth mathematical function or the infinite values. Computer operates with the finite set of the discrete objects. Moreover, the integer numbers and even the real numbers in a computer are finite sets with the finite *cardinality* N. While in mathematics real numbers are the infinite set with cardinality of the continuum $\aleph$, the integer numbers and the rational numbers are the infinite set with the cardinal number of the ordered set $\aleph_0$ [194].

Among different computational methods, the *finite difference method* is the most widely used. Such methods operate with the finite set of small elements of the continuous physical value. The method solves the differential equations by approximating the derivative by the *finite difference.* Let independent variable x be defined in the interval (X_1, X_2). Replace the continuum interval by the finite grid constructed from points $x_l = X_1 + \sum_{j=1}^{l} \Delta x_j$, where $0 \leq l \leq J$ and $\Delta x = (X_2 - X_1)/J$. An arbitrary function $f(x)$ can be replaced in the grid by finite set of values $f_l = f(x_l)$. Such description is not complete, but one can use the interpolation for $f(x)$, for example, the *linear interpolation*

$$f(x) = \frac{(x_j - x)f_{j-1} + (x - x_{j-1})f_j}{x_j - x_{j-1}}, \tag{5.1}$$

inside the interval $x \in (x_{j-1}, x_j)$. One can expect this discrete representation is appropriate for slowly variable function $f(x)$ at some selected intervals. In other words, it is not possible to describe the variation of the function $f(x)$ inside the discrete intervals [143].

The approximation of the derivative on a finite grid is illustrated with *discrete Fourier transforms.* Assume the function $f(x)$ is periodic in the interval $(X_1 = 0, X_2 = 2\pi)$ [260]. If the intervals are $\Delta x_j = \text{const} = 2\pi/J$, then the discrete Fourier transform is defined as

$$f_j = \sum_{k=1}^{J} \hat{f}_k e^{2\pi i k j/J}, \tag{5.2}$$

where the hat denotes Fourier amplitudes

$$\hat{f}_k = \frac{1}{J} \sum_{j=1}^{J} f_j e^{-2\pi i j k/J}. \tag{5.3}$$

The scalar product is defined as

$$(\mathbf{f} \cdot \mathbf{g}) \equiv \frac{1}{J} \sum_{j=1}^{J} f_j \hat{g}_j. \tag{5.4}$$

To prove expressions (5.2)–(5.3), one needs to verify the orthogonality of functions $\phi_{k,i} = e^{2\pi i k/J}$:

$$\begin{aligned}(\phi_k \cdot \phi_l) &= \frac{1}{J} \sum_{j=1}^{J} e^{2\pi i j k/J} e^{-2\pi i j l/J} = \frac{1}{J} \sum_{j=1}^{J} \left(e^{2\pi i (k-l)/J} \right)^j \\ &= \begin{cases} \frac{1}{J} e^{2\pi i (k-l)/J} \frac{1 - e^{2\pi i (k-l)}}{1 - e^{2\pi i (k-l)/J}}, & k \neq l \\ 1, k = l \end{cases} \\ &= \begin{cases} = \frac{1}{J} e^{(J+1)\pi i (k-l)/J} \frac{\sin \pi i (k-l)}{\sin(\pi i (k-l)/J)}, & k \neq l \\ 1, k = l \end{cases} = \delta_{kl}\end{aligned} \tag{5.5}$$

because the sum of *geometric progression* is $\sum_{j=1}^{J} x^j = \frac{x(1-x^N)}{1-x}$ for $x \neq 1$.

It means Fourier functions ϕ_k form the basis. The representation by *discrete Fourier transform* is exact. This implies the functional completeness [194] of functions ϕ_k. The discrete Fourier transform has a limited wavelength resolution on the finite set of numerical data f_j, $1 \leq j \leq J$, given by $\lambda_{min} = 2\pi/J$.

5.1.1 *Representation of derivatives on the grid*

One can use the expansion in *Taylor series* [113] of the function to obtain necessary derivatives of function $f(x)$ in the point x_i:

$$f(x) = \sum_{k=0}^{n} \frac{f^{(k)}(x_j)}{k!} (x - x_j). \tag{5.6}$$

To obtain n-order derivative $f_j^{(k)}$ on the grid, one needs to take into account $n+1$ points x_j from the system of linear algebraic equations in this equation. For example, for the *equidistant grid*, one has

$$f'(x_j) = \frac{f(x_{j+1}) - f(x_{j-1})}{2\Delta x} + \mathcal{O}((\Delta x)^2). \tag{5.7}$$

The grid step in power 2 in the error of the approximation $\mathcal{O}((\Delta x)^2)$ means the second *order of accuracy* of the method. In this case, it is said that formula (5.7) has second order.

Using the *discrete Fourier transform* (5.2), one can rewrite the representation (5.7) as

$$\begin{aligned} \frac{f(x_{j+1}) - f(x_{j-1})}{2\Delta x} &= \sum_{k=1}^{J} \hat{f}_k \frac{e^{ikx_{j+1}} - e^{ikx_{j-1}}}{2\Delta x} \\ &= \sum_{k=1}^{J} \hat{f}_k e^{ikx_j} \frac{2i \sin k\Delta x}{2\Delta x} \approx \sum_{k=1}^{J} \hat{f}_k e^{ikx_j} k = f'(x_j), \end{aligned} \tag{5.8}$$

if $k\Delta x \ll 1$. The numerical approximation of a *derivative on the grid* (5.7) is acceptable if $\sin k\Delta x$ is close to $k\Delta x$, or

$$k\Delta x \lesssim 1. \tag{5.9}$$

Note that it is not possible to obtain the derivative numerically of the fast oscillating function with the small wave number on the scarce grid $\Delta r > k^{-1}$.

5.2 Stability and accuracy of numerical schemes

This section introduces simplified concepts of stability and accuracy. The rigorous definitions for finite differences is presented after the classification of the types of equations in mathematical physics.

5.2.1 *Formulation of the Cauchy problem*

At this point, one can formulate the problem with the initial data for the differential (ordinary or partial) equations, the so-called *Cauchy problem*. Consider vectors defined on the *computational grid*. For the initial state $\mathbf{u}(\mathbf{r}, t_0)$, the problem to be solved is

$$\dot{\mathbf{u}}(\mathbf{r}, t) = L\mathbf{u}, \tag{5.10}$$

where L is the operator acting on the vector $\mathbf{u}$ and dot denotes differentiation with respect to time. The operator L can be nonlinear.

Example 1. Newton's law for the material point:

$$\mathbf{u} = (x, v)^T, \quad L = \begin{pmatrix} 0 & 1 \\ -F/m & 0 \end{pmatrix}, \tag{5.11}$$

where F is the force and m is the particle mass.

Example 2. Heat conduction equation in 1D $u_t - ku_{xx} = 0$, where subscripts denote partial derivatives with respect to coordinate and time. For the vector defined on the *computational grid*, one has

$$\begin{aligned} \frac{\partial}{\partial t}\mathbf{u} = L\mathbf{u}, \quad \text{where } u_j(t) &= k\frac{u_{j-1} - 2u_j + u_{j+1}}{(\Delta x)^2} \\ &= ku_{xx}(x_j, t) + \mathcal{O}((\Delta x)^2). \end{aligned} \tag{5.12}$$

The *finite difference* approximates the derivative u_{xx} in the point of grid x_j at time moment t.

The main problem in the computational physics is expressed by the question: How to formulate the *finite difference equations*? For example, one can introduce the grid for the time coordinate $t^n = t_0 + n\Delta t$ and to rewrite above equations for both cases using *explicit scheme*

$$\frac{u^{n+1} - u^n}{\Delta t} = u_t|_{t^n} + \mathcal{O}((\Delta t)^1) = L(u^n) + \mathcal{O}((\Delta t)^1). \tag{5.13}$$

In Example 1, such scheme is accurate on the fine grid with small enough parameters Δx, Δt. In Example 2, the *explicit scheme* cannot give appropriate results for any grid parameters Δx, Δt, as the numerical solution diverges for finite number of time steps. This is the problem with *stability of the scheme* [143,260].

5.2.2 *Accuracy, stability, and convergence*

The scheme is said stable if errors do not grow during integration of the evolutionary problem. Given the error ϵ^n on the nth time step, it should remain small at the next time step ϵ^{n+1} for non-growing solutions [143,260]:

$$|\epsilon^{n+1}| \leq |\epsilon^n|, \tag{5.14}$$

where the error ϵ^n is the difference of numerical solution from the exact solution of differential equation on the time step t^n. It is simple to check the errors for the linear *transfer factor* to a new time step:

$$\epsilon^{n+1} = g\epsilon^n, \tag{5.15}$$

where g is called the *transfer factor* and the stability condition requires

$$|g| \leq 1. \tag{5.16}$$

If the transfer operator to the new time step is nonlinear, it is possible to analyze the linearized transfer operator.

Consider an ordinary differential equation (ODE)

$$\frac{du}{dt} + f(u,t) = 0, \tag{5.17}$$

with the initial data $u(t^0) = u^0$. The simplest numerical scheme is the *first-order explicit Euler method*, which is

$$u^{n+1} = u^n - f(u^n, t^n)\Delta t. \tag{5.18}$$

First order here means

$$\left.\frac{u^{n+1} - u^n}{\Delta t}\right|_{t^n} = \left.\frac{du}{dt}\right|_{t^n} + \mathcal{O}((\Delta t)^1). \tag{5.19}$$

What is the stability criteria of such method? For the error ϵ^{n+1} at the new time step, one has

$$\begin{aligned} u^{n+1} + \epsilon^{n+1} &= u^n + \epsilon^n - f(u^n + \epsilon^n, t^n)\Delta t \\ &= u^n + \epsilon^n - \left(f(u^n + \epsilon^n, t^n) + \left.\frac{\partial f}{\partial u}\right|_{u^n} \epsilon^n + o(\epsilon^n)\right)\Delta t. \end{aligned} \tag{5.20}$$

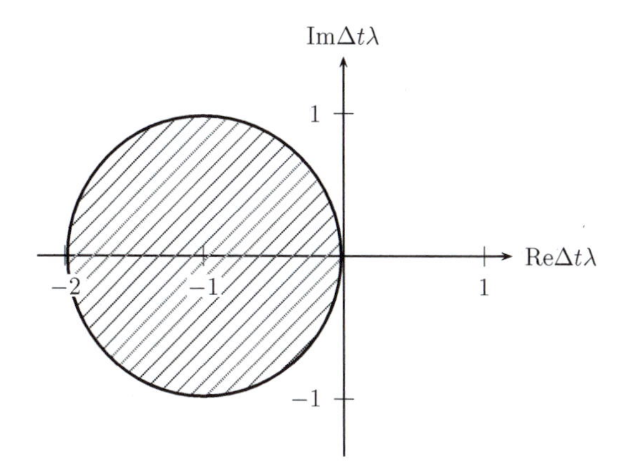

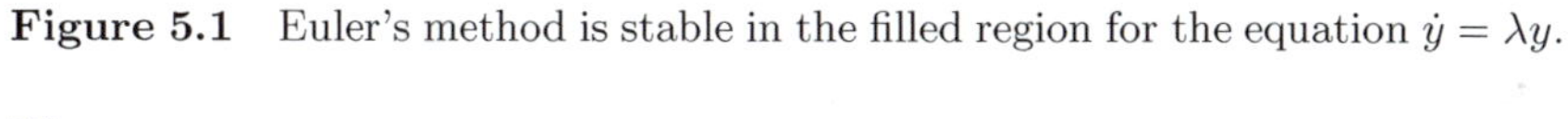
Figure 5.1 Euler's method is stable in the filled region for the equation $\dot{y} = \lambda y$.

Then

$$\epsilon^{n+1} = \left(1 - \left.\frac{\partial f}{\partial u}\right|_{u^n} \Delta t + o(\epsilon^n)\right) \epsilon^n, \tag{5.21}$$

and the *transfer factor* is

$$g = 1 - \left.\frac{\partial f}{\partial u}\right|_{n} \Delta t. \tag{5.22}$$

For the equations with the decay[3] $\partial f/\partial u < 0$, one has the following stability condition, see Fig. 5.1: The time step should be limited by the condition

$$\left|1 - \left.\frac{\partial f}{\partial u}\right|_{n} \Delta t\right| \leq 1. \tag{5.23}$$

The approximation of derivatives by *finite differences* is another requirement to the numerical scheme.

In general, the first-order scheme can give finite error for any small steps. But usually errors at every time steps have random "phase", so the first-order scheme can converge to the actual solution. But obviously the second-order scheme is preferred. The high-order scheme guarantees the converge of the solution and usually has the calculation error $\propto \Delta t^a$ and $\propto \Delta x^a$, where $a \geq 2$ [143] when the stability condition is satisfied.

[3]The growing solutions in the case $\partial f/\partial u > 0$ are useless for the analysis of the scheme.

5.3 Numerical methods for partial differential equations

In this section, numerical methods for solution of hyperbolic, elliptic, and parabolic partial differential equations are discussed.

5.3.1 *The classification of linear partial differential equations of the second order*

Recall the classification of the *second-order linear partial differential equations (PDEs)*. Consider two independent variables x, y and linear second-order differential equations [115,295,311]:

$$a_{11}u_{xx} + 2a_{12}u_{xy} + a_{22}u_{yy} + F(x, y, u, u_x, u_y) = 0, \tag{5.24}$$

where subscripts denote the derivatives with respect to variables and the coefficients a_{ij} depend on x and y. Replacing the variables as

$$\xi = \xi(x, y), \quad \eta = \eta(x, y), \tag{5.25}$$

the equation in new variables becomes

$$\alpha_{11}u_{\xi\xi} + 2\alpha_{12}u_{\xi\eta} + \alpha_{22}u_{\eta\eta} + F(x, y, u, u_x, u_y) = 0, \tag{5.26}$$

where the coefficients are

$$\alpha_{11} = a_{11}\xi_x^2 + 2a_{12}\xi_x\phi_y + a_{22}\xi_y^2, \tag{5.27}$$

$$\alpha_{12} = a_{11}\xi_x\eta_x + a_{12}(\xi_x\eta_y + \xi_y\eta_x) + a_{22}\xi_y\eta_y, \tag{5.28}$$

$$\alpha_{22} = a_{11}\eta_x^2 + 2a_{12}\eta_x\eta_y + a_{22}\eta_y^2. \tag{5.29}$$

One can obtain the relation

$$\alpha_{12}^2 - \alpha_{11}\alpha_{22} = (a_{12}^2 - a_{11}a_{22})\left[\frac{\partial(\phi, \psi)}{\partial(x, y)}\right]^2. \tag{5.30}$$

It means the sign of the discriminant $\Delta \equiv a_{12}^2 - a_{11}a_{22}$ is invariant. One can introduce the classification: When in some region D the *discriminant of the second-order PDE* is positive $\Delta > 0$, Eq. (5.24) is the *hyperbolic equation* in the region D. For $\Delta < 0$, the equation is elliptic, while for $\Delta = 0$, it is parabolic.

One can transform different types of equations to the *canonical forms of the second-order PDE*. For the *hyperbolic equation*, it is [115]

$$u_{\xi\eta} + F_1(u_\xi, u_\eta, u, \xi, \eta) = 0. \tag{5.31}$$

For the *elliptic equation*, it is

$$u_{\xi\xi} + u_{\eta\eta} + F_1(u_\xi, u_\eta, u, \xi, \eta) = 0. \tag{5.32}$$

For the *parabolic equation*, it is

$$u_{\eta\eta} + F_1(u_\xi, u_\eta, u, \xi, \eta) = 0. \tag{5.33}$$

For the linear equations, the canonical forms are as follows:

$$u_{\xi\eta} + \beta_1 u_\xi + \beta_2 u_\eta + \gamma u = f(\xi, \eta), \tag{5.34}$$

$$u_{\xi\xi} + u_{\eta\eta} + \beta_1 u_\xi + \beta_2 u_\eta + \gamma u = f(\xi, \eta), \tag{5.35}$$

$$u_{\eta\eta} + u_{\eta\eta} + \beta_1 u_\xi + \beta_2 u_\eta + \gamma u = f(\xi, \eta). \tag{5.36}$$

Consider in detail the second-order equation without mixed derivative:

$$a_{11}u_{xx} + a_{22}u_{yy} + b_1 u_x + b_2 u_y + cu = f(x, y). \tag{5.37}$$

The type of equation is defined according to the signs of the coefficients $a_{11}(x, y)$ and $a_{22}(x, y)$. When $a_{11} \neq 0$, $a_{22} \neq 0$, and the sign of $a_{11}(x, y)$ differs from the sign of $a_{22}(x, y)$, the equation is of hyperbolic type. When $a_{11} \neq 0$, $a_{22} \neq 0$, and the sign of $a_{11}(x, y)$ is the same as the sign of $a_{22}(x, y)$ and it is different from zero, the equation is of elliptic type. When $a_{11}(x, y) = 0$ or $a_{22}(x, y) = 0$, the equation is of parabolic type. For the linear equations with many independent variables $(x_1, \ldots, x_n)$:

$$\sum_i a_{ii}u_{x_i x_i} + \sum_k b_k u_{x_k} + cu = f(x_1, \ldots, x_n), \tag{5.38}$$

the following classification holds. When all coefficients are different from zero $a_{ii} \neq 0$, and all have the same sign, the equation is *elliptic*. When $a_{ii} \neq 0$, and all except for one coefficient have the same sign, the equation is *hyperbolic*. When all $a_{ii} \neq 0$ except one, and coefficients have the same sign, the equation is *parabolic*.

The famous examples of different types of equations in physics are as follows: the wave equation (hyperbolic) in 3D case

$$u_{xx} + u_{yy} + u_{zz} - a^2 u_{tt} = f(x, y, z, t), \tag{5.39}$$

the heat conduction or *diffusion equation* (parabolic)

$$u_{xx} + u_{yy} + u_{zz} - a^2 u_t = f(x, y, z, t), \tag{5.40}$$

and the *Poisson equation* (elliptic)

$$u_{xx} + u_{yy} + u_{zz} = f(x, y, z, t). \tag{5.41}$$

In Part 3, all these types of equations are encountered.

5.3.2 *Finite difference methods for the wave equation*

The type of finite difference method depends on the type of the equation. First, consider the linear *hyperbolic equation*

$$u_{tt} - a^2 u_{xx} = 0, \tag{5.42}$$

which is equivalent to the system of first-order PDEs:

$$\begin{aligned} u_t + a^2 u_x &= 0, \\ v_t + u_x &= 0. \end{aligned} \tag{5.43}$$

Changing variables to $\alpha = u + av$ and $\beta = u - av$, one has

$$\begin{aligned} \alpha_t + a\alpha_x &= 0, \\ \beta_t - a\beta_x &= 0. \end{aligned} \tag{5.44}$$

This allows one to reduce the linear second-order *hyperbolic equation* (5.42) to the first-order wave equations, like

$$u_t + au_x = 0. \tag{5.45}$$

One can write the general solution of Eq. (5.42) as

$$u = F(x - at), \tag{5.46}$$

as the wave propagating with the velocity a in the plane (x, t). Along the *characteristic* of Eq. (5.45),

$$\frac{dt}{1} = \frac{dx}{a}, \tag{5.47}$$

$x - at = \text{const}$, the function u is constant:

$$\frac{du}{dt} = \text{const.} \tag{5.48}$$

The general solution of Eq. (5.42) is the sum of two waves propagating in different directions:

$$u = F(x - at) + G(x + at), \tag{5.49}$$

where functions F and G correspond to different *characteristics* $dx/dt = \pm a$.

The *Cauchy problem* for the wave equation in the region $-\infty < x < \infty$ is the problem to find $u(x, t > 0)$ with the initial data

$$\begin{aligned} u_t + au_x &= 0, \\ u(x, 0) &= \phi(x). \end{aligned} \tag{5.50}$$

In the finite region $0 \leq x \leq l$, *Cauchy problem* also includes boundary conditions, for example, specified function u: $u(0, t) = g_0(t)$, $u(l, t) = g_l(t)$.

To solve the wave equation numerically, one can introduce the grid for the independent variables (x, t): $x_m = \Delta x m$, $0 \leq m \leq M$, $t_n = \Delta t n$, $0 \leq n$, and define the grid function u_m^n, see Fig. 5.2. One has to write *finite difference equations* which approximate the PDEs, see Fig. 5.3, for example:

$$\begin{aligned} \frac{m_m^{n+1} - u_m^n}{\Delta t} + a\frac{u_m^n - u_{m-1}^n}{\Delta x} &= f_m^n, \\ u_0^n = g_0^n, \quad 0 &\leq n, \\ u_m^0 = \phi_m, \quad 0 &\leq m \leq M. \end{aligned} \tag{5.51}$$

The grid, the *finite difference equations*, and the additional equations are referred as the *finite difference scheme*. The scheme described by Eq. (5.51) is called *explicit scheme* because one has the explicit relations for the definition of u_m^{n+1} at the new time step.

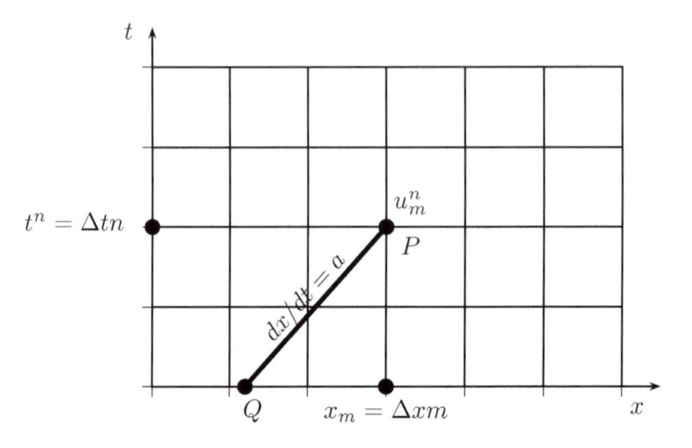

Figure 5.2 Example of the *computational grid*. The bold curve $dx/dt = a$ is the *characteristic* for the *hyperbolic equation*.

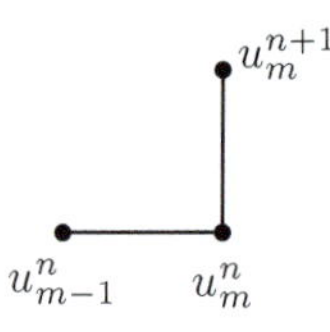

Figure 5.3 The upwind asymmetric explicit template for the *hyperbolic equation*.

5.3.2.1 *The accuracy of the finite difference scheme*

In the finite difference scheme

$$\frac{\partial u}{\partial t}(x_m, t^n) = \frac{u_m^{n+1} - u_m^n}{\Delta t} + \mathcal{O}(\Delta t),$$

$$\frac{\partial u}{\partial x}(x_m, t^n) = \frac{u_m^n - u_{m-1}^n}{\Delta x} + \mathcal{O}(\Delta x), \tag{5.52}$$

the *finite difference equations* approximate the first-order PDE for the grid intervals Δt, Δx.

Other types of the schemes are possible:

- The explicit asymmetric *forward difference scheme*, see Fig. 5.4,

$$\frac{u_m^{n+1} - u_m^n}{\Delta t} + a\frac{u_{m+1}^n - u_m^n}{\Delta x} = f_m^n \tag{5.53}$$

  has the error of the approximation $\mathcal{O}(\Delta t) + \mathcal{O}(\Delta x)$.

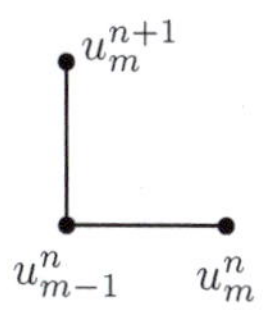

Figure 5.4 The forward difference explicit template for the *hyperbolic equation.*

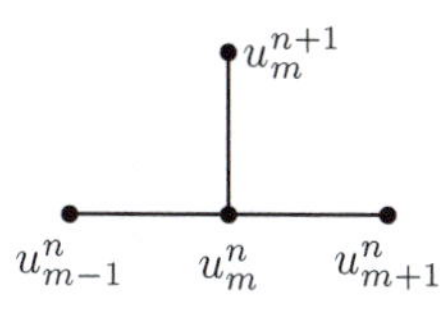

Figure 5.5 The symmetric explicit template.

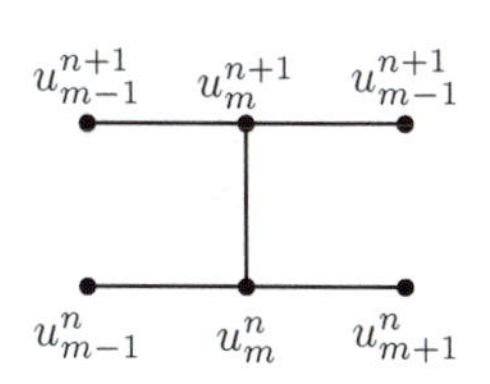

Figure 5.6 The implicit second-order template.

- The explicit symmetric scheme, see Fig. 5.5,

$$\frac{u_m^{n+1} - u_m^n}{\Delta t} + a\frac{u_{m+1}^n - u_{m-1}^n}{2\Delta x} = f_m^n \tag{5.54}$$

  has the error of the approximation $\mathcal{O}(\Delta t) + \mathcal{O}((\Delta x)^2)$.
- The implicit *central difference* scheme, see Fig. 5.6,

$$\frac{u_m^{n+1} - u_m^n}{\Delta t} + \frac{a}{2}\frac{u_{m+1}^n - u_{m-1}^n}{2\Delta x} + \frac{a}{2}\frac{u_{m+1}^{n+1} - u_{m-1}^{n+1}}{2\Delta x} = f_m^n \tag{5.55}$$

  has the error of the approximation $\mathcal{O}((\Delta t)^2) + \mathcal{O}((\Delta x)^2)$. However, it requires to solve the set of the linear algebraic equations to find u_m^{n+1} at the new time step t^{n+1}. Such scheme is called *implicit scheme.*[4]

[4] Implicit schemes were invented in the struggle to avoid problems with numerical solutions, emerged in the Soviet Atomic project. This fact was communicated to one of us (GV) by Isaak Khalatnikov, see also Ref. [190], p. 49.

5.3.2.2 *Characteristics and the stability*

For the wave equation in more general form

$$u_t + au_x = f(x,t), \tag{5.56}$$

one has the relation

$$\frac{dt}{1} = \frac{dx}{a} = \frac{du}{f}, \tag{5.57}$$

or along the *characteristic* $\frac{dx}{dt} = a$ [295],

$$\frac{du}{dt} = f. \tag{5.58}$$

The unknown function u at the point $P = (t,x)$ depends on the region Q on the *characteristic* $x - at = \text{const}$. The region Q is called the *domain of dependence* of the solution in the point P: $u(P) = \phi(Q)$, see Fig. 5.2.

One can predict the stability for the *explicit schemes* for *hyperbolic equations.* When the domain of dependence of the *finite difference equation* contains the domain of dependence of the PDE, the scheme can be stable. On the contrary, when the domain of dependence of the *finite difference equation* does not contain the domain of dependence of the PDE, the scheme is unstable. For instance, the explicit *upwind* (the integration against the flow) scheme (5.51) illustrated in Fig. 5.3 is stable for small enough time step $\Delta t \leq a\Delta x$ (u_m^{n+1} depending on the set of points u_{m-1}^n, u_m^n), while the *forward difference scheme* given by Eq. (5.53) and illustrated in Fig. 5.4 is unstable. The numerical solution of the explicit symmetric scheme given by Eq. (5.54) and Fig. 5.5 does not include the *domain of dependence*, while u_m^{n+1} should depend on u_m^n. The characteristic properties and the stability of the *implicit scheme* (5.55), shown in Fig. 5.6 are not obvious: u_m^{n+1} depends on the set of points u_{m-1}^{n+1}, u_{m+1}^{n+1} u_{m-1}^n, u_m^n, u_{m+1}^n. Also the stability of the *explicit scheme* (5.54) shown in Fig. 5.4 is not obvious. Both schemes at Figs. 5.4 and at 5.5 include the *domain of dependence.* Such schemes are stable, but the *explicit scheme* (5.54) is stable only for small time steps Δt.

5.3.2.3 *Von Neumann spectral method and the stability*

In numerical analysis, a *von Neumann stability analysis* also known as *Fourier stability analysis* is a procedure used to verify the stability of *finite difference schemes* applied to linear PDEs [97,116,312]. One can consider the evolutionary scheme as the linear operator A:

$$U^{n+1} = AU^n. \tag{5.59}$$

For stable scheme, the norm of the operator

$$||A|| \equiv \sup_U \frac{||AU||}{||U||} \tag{5.60}$$

should be limited

$$||A^n|| \leq C, \quad n\Delta t \leq T, \tag{5.61}$$

or

$$||A|| \leq 1 + C\Delta t, \tag{5.62}$$

where C is a constant. One can estimate the norm of the operator $||A||$ as the maximal eigenvalue λ:

$$AU = \lambda U \tag{5.63}$$

because

$$\max |\lambda| \leq ||A||. \tag{5.64}$$

The expected necessary condition for the stability is

$$|\lambda| \leq C, \quad n\Delta t \leq T, \text{ or } |\lambda| \leq 1 + C\Delta t. \tag{5.65}$$

If the eigenvectors constitute the basis, the condition for the eigenvalues (5.65) is sufficient for the stability.

The numerical solution on the infinite plane $\infty < m < \infty$ can be presented as

$$u_m = \frac{1}{\sqrt{2\pi}} \int_0^{2\pi} W(\phi) e^{im\phi} d\phi, \quad \text{where } W(\phi) = \frac{1}{\sqrt{2\pi}} \sum u_m e^{-im\phi}, \tag{5.66}$$

and

$$v_m = e^{im\phi} \tag{5.67}$$

are eigenfunctions of the operator A. One can use *Parseval's identity* [194]

$$\sum_m |u_m|^2 = \int_0^{2\pi} |W(\phi)|^2 d\phi. \tag{5.68}$$

Then one can write

$$Au_m = \frac{1}{\sqrt{2\pi}} \int W(\phi) A v_m d\phi = \frac{1}{\sqrt{2\pi}} \int W(\phi) \lambda e^{im\phi} d\phi. \tag{5.69}$$

Parseval's identity gives

$$\begin{aligned} \sum |Au_m|^2 &= \int_0^{2\pi} |W(\phi)|^2 |\lambda|^2 d\phi \le (1 + \mathcal{O}(\Delta t)) \int_0^{2\pi} |W(\phi)|^2 d\phi \\ &= (1 + \mathcal{O}(\Delta t)) \sum_m |u_m|^2. \end{aligned} \tag{5.70}$$

Relations $|\lambda| \le 1 + \mathcal{O}(\Delta t)$ and $||A|| \le 1 + \mathcal{O}(\Delta t)$ are equivalent.

Then the necessary and sufficient conditions for the stability are the limited eigenvalue

$$|\lambda(\phi)| \le 1 + \mathcal{O}(\Delta t) \tag{5.71}$$

for the eigenfunction (5.67) for any ϕ. To define the eigenvalue $\lambda(\phi)$ of the scheme, one can search for the solution in the form

$$u_m^n = \lambda^n e^{im\phi}. \tag{5.72}$$

Now one can analyze the stability of the *explicit schemes* (5.51), (5.53), (5.54), and the *implicit scheme* (5.55) for the *hyperbolic equation*. For the scheme (5.51) of Fig. 5.3, one has the equation

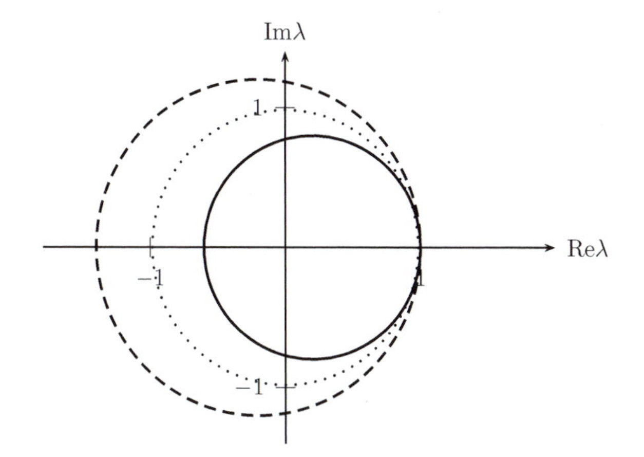

Figure 5.7 The spectra $\lambda(\phi)$ for the *upwind* asymmetric *explicit scheme* (5.51): $0 < a\Delta t/\Delta x < 1$ (solid curve), $1 < a\Delta t/\Delta x$ (dashed curve), and $|\lambda| = 1$ (dotted curve).

for the eigenvalue (Fig. 5.7)

$$\lambda(\phi) = 1 + \frac{a\Delta t}{\Delta x}(e^{-i\phi} - 1). \tag{5.73}$$

On the plane with coordinates $(\mathrm{Re}\lambda, \mathrm{Im}\lambda)$, the eigenvalue $\lambda(\phi)$ is the circle of the radius $a\Delta t/\Delta x$ with the center $(1 - a\Delta t/\Delta x, 0)$. Obviously, $|\lambda(\phi)| \leq 1$ if $a\Delta t/\Delta x \leq 1$ for the positive value $a \geq 0$. This condition is expected for scheme (5.51) from the characteristic properties of the *hyperbolic equation.*

The explicit difference scheme with the forward difference (5.53) gives

$$\lambda(\phi) = 1 - \frac{a\Delta t}{\Delta x}(e^{i\phi} - 1). \tag{5.74}$$

For $\phi = \pi$ $\lambda = 1 + 2a\Delta t/\Delta x$ and $|\lambda| > 1 + \mathcal{O}(\Delta t)$, see Fig. 5.8. This implies that the scheme is unstable.

For the *explicit scheme* with the *central difference* (5.54), one has

$$\lambda(\phi) = 1 - i\frac{a\Delta t}{\Delta x}\sin\phi, \tag{5.75}$$

and $|\lambda(\phi)| = 1 + (a\Delta t/\Delta x)^2 \sin^2\phi$ (Fig. 5.9). In practice, the scheme is useless because it is stable $|\lambda(\phi)| \leq 1 + \mathcal{O}(\Delta t)$ only for very small time steps $\Delta t = \mathcal{O}((\Delta x)^2)$.

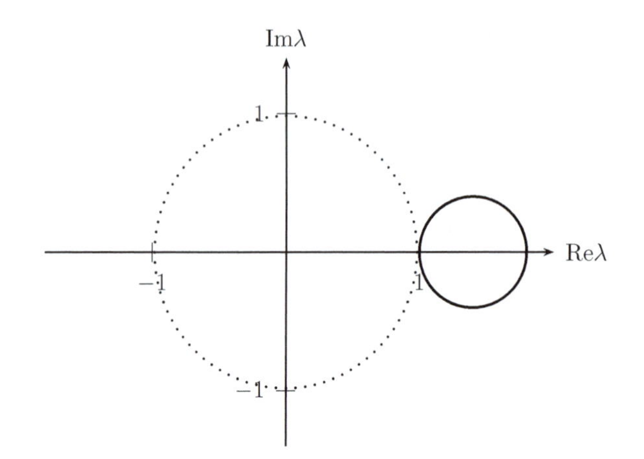

Figure 5.8 The spectra $\lambda(\phi)$ for the forward unstable *explicit scheme* (5.53): for any time steps $\Delta t > 0$, the solid curve is outside the region $|\lambda| = 1$ (dotted curve).

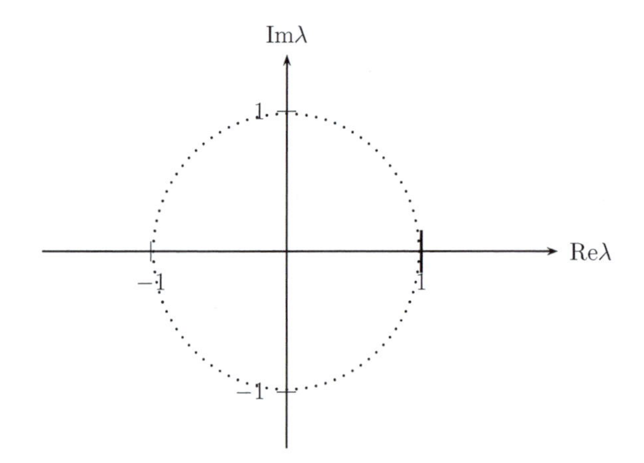

Figure 5.9 The spectra $\lambda(\phi)$ for the symmetric *explicit scheme* (5.54) is the solid curve near $\mathrm{Re}\lambda = 1$, and $|\lambda| = 1$ (dotted curve).

For the *implicit scheme* (5.55), one has

$$\lambda(\phi) = \frac{1 - i\frac{a\Delta t}{2\Delta x}\sin\phi}{1 + i\frac{a\Delta t}{2\Delta x}\sin\phi}, \tag{5.76}$$

and $|\lambda|^2 = 1$. The *implicit scheme* is stable for any time step and also has the second-order accuracy. However, the second-order accuracy is not always useful for the *hyperbolic equation* because its monotonic property.

The *spectral method* for the investigation of the stability of *finite difference schemes* is more useful. In the case of nonlinear equations, one can try to investigate the stability of linearized equations with "frozen" coefficients. The next step is investigation of the stability and the convergence of the scheme in the numerical experiment. Other methods for investigations of the stability can be found in the literature, for example, the energetic method [267].

5.3.2.4 *Monotonic property for the hyperbolic equation and high-order schemes*

The scheme is said to be *monotonic* if it produces the *monotonic solution* from the monotonic state [141,142]. This property of the scheme is important for the *hyperbolic equations*. Such hyperbolic equations describe, for example, hydrodynamics of ideal gas. In the first approximation, hydrodynamic equations can be considered as a set of the wave equations for density, velocity, and *pressure*. In the case of strong jumps (*shock wave* (SW) and *contact discontinuity*), it is important to avoid artificial oscillation near such a jump. For non-monotonic scheme, the value of density and *pressure* on the grid can even become negative and consequently unphysical.

Consider the *explicit scheme* for the wave equation

$$u_m^{n+1} = \sum_k c_k u_{m+k}^n \tag{5.77}$$

with constant coefficients c_k. The criteria of the *monotonic scheme* is the condition for all coefficients

$$c_k \geq 0. \tag{5.78}$$

The proof is the following. Write the following difference:

$$u_{m+1}^{n+1} - u_m^{n+1} = \sum_k c_k (u_{m+1+k}^n - u_{m+k}^n). \tag{5.79}$$

Obviously, if the initial state at t^n is monotonic, for example, all $u_{m+1+k}^n - u_{m+k}^n \leq 0$, then the solution at new time step t^{n+1} is monotonic either in the selected example $u_{m+1+k}^{n+1} - u_{m+k}^{n+1} \leq 0$ holds as well. Proof by contradiction is as follows. Let the coefficient be negative, $c_{k_0} < 0$. Then if one takes

$$u_k^n = \begin{cases} 1, & k \le k_0, \\ 0, & k > k_0, \end{cases} \tag{5.80}$$

the result is $u_{m+1}^{n+1} - u_m^{n+1} = \sum_k c_k(u_{m+1+k}^n - u_{m+k}^n) = c_{k_0-(m+1)}$, and $u_{-1}^{n+1} - u_{-2}^{n+1} = c_{k_0} < 0$. It implies that the scheme is not monotonic.

It is interesting to stress two important properties of scheme (5.77). The first, obvious property is that the *implicit scheme*

$$\sum_k a_k u_m^{n+1} = \sum_k b_k u_{m+k}^n \tag{5.81}$$

can be reduced to the *explicit scheme* (5.77) by the explicit solution of the system of linear equations. Another interesting property of scheme (5.77) is its stability. When to the monotonic property $c_k \ge 0$, one adds the natural requirement $\sum_k c_k = 1$:

$$\max_m |u_m^{n+1}| \le \sum_k c_k \max_m |u_m^n| \le \max_m |u_m^n|. \tag{5.82}$$

one new constant solution appears $u_m^{n+1} = \text{const}$ if initially $u_m^n = \text{const}$.

The absence of the monotonic second-order scheme for the wave equation is another important property [142]. Consider again the general scheme (5.77). One can select the parabola

$$u(x,0) = \left(\frac{x}{\Delta x} - \frac{1}{2}\right)^2 - \frac{1}{4}, \tag{5.83}$$

where h is the grid size, as an initial state. The solution of the wave equation is a shifted parabola

$$u(x,t) = \left(\frac{x+t}{\Delta x} - \frac{1}{2}\right)^2 - \frac{1}{4}. \tag{5.84}$$

The second-order (or higher order) scheme has the error $\mathcal{O}((\Delta x)^2 + (\Delta t)^2)$, and it will reproduce this parabola in the numerical solution.

The initial finite difference function if positive

$$u_m^0 = \left(m - \frac{1}{2}\right)^2 - \frac{1}{4} \geq 0, \tag{5.85}$$

but the second-order scheme gives

$$u_m^n = \left(m + \frac{t}{\Delta x} - \frac{1}{2}\right)^2 - \frac{1}{4}. \tag{5.86}$$

For $m = -1$ and $t^n = \Delta x$, the numerical solution becomes negative

$$u_{-1}^n = -\frac{1}{4}. \tag{5.87}$$

But the *monotonic scheme* (5.77) should have coefficients $c_k \geq 0$ and give all $u_m^n \geq 0$. Then one arrives to an important conclusion. In general case, it is useful to construct high-order resolution schemes for *hyperbolic equations* only if the flow is smooth. It is necessary to reduce the order of the scheme to the first in the region with high gradients or discontinuities.

5.3.3 *Finite difference methods for the parabolic equation*

Consider now the *parabolic equation*

$$u_t = a^2 u_{xx} \tag{5.88}$$

and adopt the *explicit scheme* represented in Fig. 5.5

$$\frac{u_m^{n+1} - u_m^n}{\Delta t} = a^2 \frac{u_{m-1}^n - u_m^n + u_{m+1}^n}{(\Delta x)^2}, \tag{5.89}$$

having the error of the approximation $\mathcal{O}(\Delta t + (\Delta x)^2)$. The test solution

$$v_m^n = \lambda^n e^{im\phi} \tag{5.90}$$

gives

$$\lambda(\phi) - 1 = 2\frac{a^2 \Delta t}{(\Delta x)^2}(\cos\phi - 1) = -4\frac{a^2 \Delta t}{(\Delta x)^2}\sin^2\frac{\phi}{2}. \tag{5.91}$$

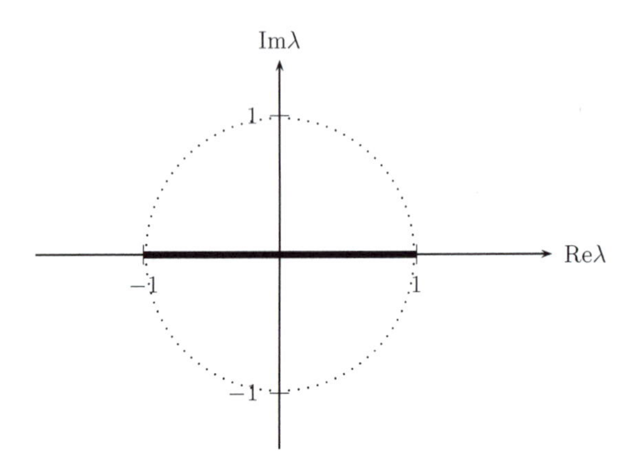

Figure 5.10 The spectra $\lambda(\phi)$ for the symmetric *explicit scheme* (5.89) (solid curve), and $|\lambda| = 1$ (dotted curve) for $\frac{a^2\Delta t}{(\Delta x)^2} = \frac{1}{2}$.

The *explicit scheme* is stable, see Fig. 5.10, only for small enough time step given by

$$\frac{a^2\Delta t}{(\Delta x)^2} \leq \frac{1}{2}. \tag{5.92}$$

More adequate for *parabolic equation* is the implicit second-order $\mathcal{O}((\Delta t)^2 + (\Delta x)^2)$ scheme shown in Fig. 5.6 [116],

$$\frac{u_m^{n+1} - u_m^n}{\Delta t} = a^2 \frac{u_{m-1}^n - u_m^n + u_{m+1}^n + u_{m-1}^{n+1} - u_m^{n+1} + u_{m+1}^{n+1}}{2(\Delta x)^2}, \tag{5.93}$$

that gives the coefficient λ as

$$\lambda(\phi) = \frac{1 - 2\frac{a^2\Delta t}{(\Delta x)^2}\sin^2\frac{\phi}{2}}{1 + 2\frac{a^2\Delta t}{(\Delta x)^2}\sin^2\frac{\phi}{2}}. \tag{5.94}$$

Such a scheme is stable $|\lambda(\phi)| \leq 1$ for any time step.

5.3.4 *Finite difference methods for the elliptic equation*

Consider the system

$$\Delta u + f = 0, \tag{5.95}$$

where Δ is the Laplace operator, with the boundary condition

$$u|_{\text{bound}} = \phi. \tag{5.96}$$

One can solve the evolutionary problem for the *parabolic equation*

$$v_t = \Delta v + f, \tag{5.97}$$

$$v|_{\text{bound}} = \phi, \tag{5.98}$$

with some arbitrary initial value $v(t = 0) = v_0$ until the stationary state is obtained. For the difference $w = v - u$, one has

$$w_t = \Delta w, \tag{5.99}$$

$$w|_{\text{bound}} = 0, \tag{5.100}$$

$w(t = 0) = v_0 - u$, and

$$w \propto e^{-\lambda t}, \tag{5.101}$$

where λ is the minimal eigenvalue of the operator Δ.

Consider the following example

$$u_t + \Delta u = u_t + u_{xx} + u_{yy} = 0 \tag{5.102}$$

in the region $0 \le x \le \pi$, $0 \le y \le \pi$ with the boundary condition

$$u|_{\text{bound}} = 0. \tag{5.103}$$

Write down the *explicit scheme*

$$\frac{U^{n+1} - U^n}{\Delta t} = \Lambda_{11}U^n + \Lambda_{22}U^n, \tag{5.104}$$

where the operators Λ_{11}, Λ_{22} are

$$\begin{aligned} \Lambda_{11}U &= \frac{u_{l+1,m} - 2u_{lm} + u_{l-1,m}}{\Delta x_1^2}, \\ \Lambda_{22}U &= \frac{u_{l,m+1} - 2u_{lm} + u_{l,m+1}}{\Delta x_2^2}, \end{aligned} \tag{5.105}$$

and

$$U^{n+1} = AU^n, \quad \text{where} \quad A = I + \Delta t(\Lambda_{11} + \Lambda_{22}). \tag{5.106}$$

The solution can be searched in the form

$$U \sim e^{il\phi}e^{im\psi} \tag{5.107}$$

with $u_{lm} = 0$ on the boundary for $l, m = 0, L$. The eigenvalues are $-a_{11} = -\frac{\Delta t}{(\Delta x)^2}\sin^2\frac{\phi}{2}$ for the eigenfunctions $e^{\pm il\phi}$ (the corresponding real eigenfunction $\sin(l\phi) = \frac{e^{il\phi}-e^{-il\phi}}{2i})$ of the operator $\Delta t\Lambda_{11}$ and $a_{22} = -\frac{\Delta t}{(\Delta x)^2}\sin^2\frac{\psi}{2}$ for the eigenfunction $\sin(m\psi)$ of the operator $\Delta t\Lambda_{22}$. For the operator $A = I + \Delta t(\Lambda_{11}+\Lambda_{22})$, one has the eigenfunctions

$$u_{lm}^{k_1k_2} = \sin(k_1lh)\sin(k_2mh), \tag{5.108}$$

with $1 \leq k_1 \leq L-1$ and $1 \leq k_2 \leq L-1$ and eigenvalues

$$\begin{aligned}\rho(\Delta t, k_1, k_2) &= 1 - a_{11}(k_1) - a_{22}(k_2)\\ &= 1 - \frac{\tau}{(\Delta x)^2}\sin^2\frac{k_1\Delta x}{2} - \frac{\Delta t}{(\Delta x)^2}\sin^2\frac{k_2\Delta x}{2}.\end{aligned} \tag{5.109}$$

One can approximate the norm of the operator for a small h:

$$\begin{aligned}||A(\Delta t)|| &= \max_{k_1,k_2}|\rho(\Delta t, k_1, k_2)| = |1-2a(1)|\\ &= |1 - 2\cdot 4\frac{\Delta t}{(\Delta x)^2}\sin^2\frac{\Delta x}{2}| \approx |1-2\Delta t|.\end{aligned} \tag{5.110}$$

The stability of the *explicit scheme* requires $\frac{\Delta t}{(\Delta x)^2} \leq \frac{1}{2}$, and $||A|| = 1-(\Delta x)^2/2$. One can find the number of time steps to obtain the small enough numerical solution $||U^N|| \leq \epsilon||U^0||$ for any small ϵ: $(1-(\Delta x)^2/2)^N \leq \epsilon$. It implies that the required number of the steps in the explicit *relaxation* scheme is proportional to the large value $N \geq -L^2\frac{2}{\pi}\ln\epsilon \propto L^2$. For this *explicit scheme*, there exists the so-called *Chebyshev set of variable time steps* Δt to achieve sufficient accuracy with reduced number of time steps [143]. For the *implicit scheme* with alternate directions (the time step consists of two time steps with the integration along only one direction), the number of time steps is small $N \geq -L\frac{1}{2\pi}\ln\epsilon$.

Alternative approaches for the *elliptic equation* are the solution for the system of linear algebraic equations arising after the replacement of the derivatives by the *finite differences* for the grid functions or implementation of *Fourier transforms* in the case of the periodic boundary conditions.

5.3.5 *Finite difference methods for multidimensional problems and nonlinear equations*

So far, 1D *finite difference schemes* for different types of equations of the mathematical physics were considered.

Often, numerical solution for multidimensional problem is required. The multidimensional problem can be split onto the set of 1D problems for the application of a *finite difference method.* It is the so-called *dimensional splitting* method. For the 3D case, for example, at every time step, one can consider three separate problems taking into account the derivatives along one direction [143]. This approach is effective and stable and has satisfactory accuracy. This approach can be especially useful for *implicit schemes*, for example, *Crank–Nicolson method* for the heat conduction [116].

Another important problem is the *stability of the scheme* for nonlinear differential equations. To study the stability, one can consider the linearized problems with the assumption of the constant coefficients in front of derivatives. The stability of the problem with such "frozen" coefficients can be verified by means of the *spectral method* [143]. The *stability of the scheme* for the nonlinear problem can be verified by the computational experiment.

5.3.6 *Fast Fourier transform*

The *discrete Fourier transforms* for the discrete function f_i and the Fourier amplitudes $\hat{f}_k$ were introduced in Eq. (5.2). Writing $J = 2J_1$ it is possible to reduce the calculations of Fourier transform to the Fourier transforms on the two subgrids:

$$u_j = \sum_{k=1}^{J_1} \hat{u}_k e^{2\pi ikj/J_1} = \sum_{k=1}^{J_1} \hat{f}_{2k} e^{2\pi i2kj/J_1}, \quad \hat{u}_k = \hat{f}_{2k},$$

$$v_j = \sum_{k=1}^{J_1} \hat{v}_k e^{2\pi ikj/J_1} = \sum_{k=1}^{J_1} \hat{f}_{2k+1} e^{2\pi i(2k+1)j/N_1} e^{-2\pi ij/J_1},$$

$$\hat{v}_k = \hat{f}_{2k+1}, \quad j = 1, \ldots, J_1. \tag{5.111}$$

One obtains a set of relations

$$f_j = u_j + e^{-2\pi ij/J_1} v_j, \quad j = 1, \ldots, J_1. \tag{5.112}$$

Analogously, one can obtain the following set of transformations:

$$f_{J_1+j} = u_j - e^{-2\pi ij/J_1} v_j, \quad j = 1, \ldots, J_1 - 1. \tag{5.113}$$

The calculations on the subgrid require the number of operations $\sim J_1 = J/2$, so the number of operations on the grid J is $T(J) \leq 2T(J/2) + 2J$. The optimal grid from the viewpoint of the minimization of the calculations is the power of 2: $J = 2^s$, where s is a natural number. Then the *fast Fourier transform* allows one to reduce the number of calculations from J^2 to $\sum_{k=1}^{J} 2^k T(J/2^k) \propto 2J \log_2 J$.

5.3.7 *Variational methods*

Consider the linear self-adjoint positively defined operator L with the following properties: the scalar product $(Lu \cdot v) = (u \cdot Lv)$ and $(Lv \cdot v) \geq 0$. In Euclidean space, the matrix corresponding to the self-adjoint operator is symmetric $A = A^T$. The following problem

$$Lu^* = f, \tag{5.114}$$

and the variational problem

$$I[u^*] = \min_u I[U], \text{ where } I[u] = (Lu, u) - 2(u, f) \tag{5.115}$$

are equivalent. In order to proof this statement, one can evaluate the following expression with the new parameter t and any function v: $\frac{d}{dt} I[u^* + tv]\big|_{t=0} = 0$. One has

$$I[u^* + tv] = (Lu^* + tLv, u^* + tv) - 2(u^*, f) - 2t(v, f), \tag{5.116}$$

and $\frac{d}{dt} I[u^* + tv]\big|_{t=0} = (Lv, u^*) + (Lu^*, v) - 2(v, f) = 2((Lu^*, v) - (v, f)) = 2(Lu^* - f, v) = 0$. Since the function v is arbitrary, the proof that the assumption (5.115) gives the relation (5.114) is completed. To prove the relation (5.115) from the relation (5.114), one can evaluate the expression $I[u^* + v] = (Lu^*, u^*) - 2(u^*, f) + (Lu^*, v) + (Lv, u^*) - 2(v, f) + (Lv, v) = (Lu^*, u^*) + (Lv, v)$. Because $(Lv, v) \leq 0$, one obtains $I[u^* + v] \leq I[u^*]$.

The *Dirichlet problem* for the *Laplace equation* $\nabla u = f$ in the region with the given value on the boundary $u|_B = u_0$ is the appropriate example for the variational problem. The *variational method* in the numerical simulation can be useful for the suitable selection of the base functions and the representation of the solution as the linear combination of the base functions with coefficients to be defined. The *Ritz–Galerkin variational method* converts a continuous operator problem, such as a differential equation, to a discrete problem, see examples in Refs. [143,292].

5.4 The method of lines

The *method of lines* is a method for solving PDEs in which all but 1D are discretized [73,274]. Obviously, if one has the evolutionary problem (*Cauchy problem*), it is possible to introduce the grid in the space and replace all partial derivatives (except time derivatives) by *finite differences* of the grid functions. The necessary integrals in the *phase space* are replaced by sums. After this procedure, one has the system of ODEs describing the evolution of grid functions in time. Then one can apply the standard methods for the solution of ODE systems. The realization of the method of lines for Boltzmann equations is discussed in Chapter 6.

5.5 ODE systems and methods of their solutions

Consider the initial value problem $\mathbf{y}(0) = \mathbf{y}_0$ for the system of ODEs:

$$\dot{\mathbf{y}} = \mathbf{f}(\mathbf{y}, t). \tag{5.117}$$

In the beginning of this chapter, the *first-order explicit Euler method* (5.18) was considered as the simplest illustration for the solution of the problem (5.117). The condition for the convergence of the numerical solution to the mathematical solution was formulated: the accuracy and the stability of the method. The expansion in *Taylor series* gives the accuracy of the method. The stability can be verified by the test equation $\dot{y} = \lambda y$. The scheme is stable if the errors of the approximation do not grow.

The function $\mathbf{f}(\mathbf{y}, t)$ should satisfy the *Lipschitz condition* [157]: There exists a constant L such that

$$||\mathbf{f}(\mathbf{y}, t) - \mathbf{f}(\mathbf{y}^*, t)|| \leq L||\mathbf{y} - \mathbf{y}^*||. \tag{5.118}$$

If this condition is satisfied, the initial value problem has the unique solution. For the *explicit one-step Euler method*

$$\mathbf{y}^{n+1} = \mathbf{y}^n + \Delta t \mathbf{f}(\mathbf{y}^n, t^n) \tag{5.119}$$

with the time step $\Delta t = t^n/n$, the definition of the convergence is

$$\mathbf{y}^n \longrightarrow \mathbf{y}(t) \tag{5.120}$$

for all $0 \leq t \leq b$ as $n \longrightarrow \infty$ and $\mathbf{y}_0 \longrightarrow \mathbf{y}(0)$. Consider the stability of numerical method for the special model problem with the complex value λ

$$\dot{y} = \lambda y. \tag{5.121}$$

The reason for consideration of such a simple problem is the following. For the system of ODEs (5.117), one can consider the local linearized problem

$$\dot{\mathbf{y}} = \frac{\partial \mathbf{f}}{\partial \mathbf{y}} \mathbf{y} \tag{5.122}$$

with a constant *Jacobi matrix* $J = \frac{\partial \mathbf{f}}{\partial \mathbf{y}}$. By the definition of right eigenvectors $J\mathbf{r}_j = \lambda_j \mathbf{r}_j$ and $\Lambda R = \mathrm{diag}(\lambda_1, \ldots, \lambda_n)R$, $R^{-1}JR = \Lambda$, where matrix R consists of right eigenvectors $R = (\mathbf{r}_1, \ldots, \mathbf{r}_n)^T$, and $\Lambda = \mathrm{diag}(\lambda_1, \ldots, \lambda_n)$. If the *Jacobi matrix* J is non-degenerate (all eigenvalues are different from zero), the system can be transferred to the diagonal form:

$$\frac{d}{dt}(R^{-1}\mathbf{y}) = \Lambda(R^{-1}\mathbf{y}), \tag{5.123}$$

so one can investigate the stability of the method for one ordinary differential equation.

As an example, consider the widely used the Classical *Runge–Kutta method*, developed around 1900 by the German mathematicians C. Runge and M. W. Kutta [196]. It is a four-stage *explicit method* of fourth order:

$$k_0 = \Delta t f(y^n, t^n),$$

$$k_1 = \Delta t f\left(y^n + \frac{1}{2}k_0, t^n + \Delta t/2\right),$$

$$k_2 = \Delta t f\left(y^n + \frac{1}{2}k_1, t^n + \Delta t/2\right),$$
$$k_3 = \Delta t f(y^n + k_2, t^{n+1}),$$
$$y^{n+1} = y^n + \frac{1}{6}(k_0 + 2k_1 + 2k_2 + k_3). \tag{5.124}$$

This method can be viewed [157] as an attempt to extend the *Simpson quadrature*:

$$\int_{t^n}^{t^{n+1}} f(t)dt \approx \frac{\Delta t}{6}\left(f(t^n) + 4f\left(t^n + \frac{\Delta t}{2} + f(t^{n+1})\right)\right). \tag{5.125}$$

If $f(y,t)$ is a function of t only, the Runge–Kutta formula and Simpson's integral are equivalent [157].

To verify the order of the method, one can use the expansion in *Taylor series* around t^n with the accuracy $\mathcal{O}((\Delta t)^4)$.

For the investigation of the stability, one can examine the formula for the equation $\dot{y} = \lambda y$. One has

$$k_1 = \Delta t\lambda\left(1 + \frac{1}{2}\Delta t\lambda\right)y^n,$$
$$k_2 = \Delta t\lambda\left(1 + \frac{1}{2}\Delta t\lambda + \frac{1}{2}\Delta t\lambda\left(1 + \frac{1}{2}\Delta t\lambda\right)\right)y^n,$$
$$k_3 = \Delta t\lambda\left(1 + y^n + \Delta t\lambda\left(1 + \frac{1}{2}\Delta t\lambda + \frac{1}{2}\Delta t\lambda\left(1 + \frac{1}{2}\Delta t\lambda\right)\right)\right)y^n,$$
$$y^{n+1} = \left(1 + \Delta t\lambda + \frac{1}{2}(\Delta t\lambda)^2 + \frac{1}{6}(\Delta t\lambda)^3 + \frac{1}{24}(\Delta t\lambda)^4\right)y^n. \tag{5.126}$$

In the accordance with the fourth-order approximation, $y^{n+1} = y^n \sum_{n=0}^{4} \frac{(\Delta t\lambda)^4}{n!}$. Thus, the region of the stability is that area in which the *growth factor* is limited

$$\left|\sum_{n=0}^{4} \frac{(\Delta t\lambda)^4}{n!}\right| < 1. \tag{5.127}$$

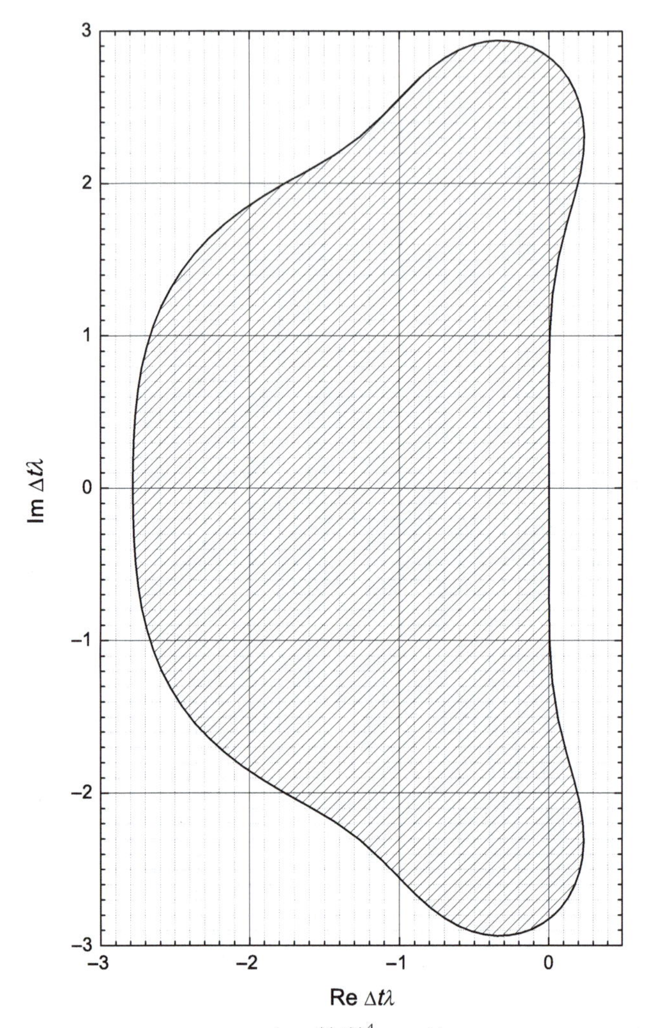

Figure 5.11 The filled region $\sum_{n=0}^{4} \frac{(\Delta t\lambda)^4}{n!} \leq e^{i\phi}$ is the region of stability for the explicit fourth-order *Runge–Kutta method*.

The region of the stability in the complex plane for coordinates $(\mathrm{Re}(\Delta t\lambda), \mathrm{Im}(\Delta t\lambda))$ limited by the curve $\sum_{n=0}^{4} \frac{(\Delta t\lambda)^4}{n!} = e^{i\phi}$ ($|e^{i\phi}|= 1$) is shown in Fig. 5.11. For the *explicit method*, time steps are limited to $-2.8 \lesssim \mathrm{Re}(\Delta t\lambda) \lesssim 0$ for $\mathrm{Im}(\Delta t\lambda) = 0$. For the *implicit method*, it is possible to have the absolutely stable *A-stable method* for any $\mathrm{Re}(\Delta t\lambda) \leq 0$.

5.6 Stiff systems and Gear's method

One can approximate the *Cauchy problem* (5.117) near the exact solution $y(t)$,

$$\dot{\mathbf{y}} - J\mathbf{y} - f(\mathbf{y}, t) = 0, \tag{5.128}$$

with *Jacobi matrix* $J(t)$. If the variation of $J(t)$ is small in the vicinity of fixed t, the solution can be represented in this vicinity as

$$\mathbf{y} \approx \mathbf{y}(t) + \sum_i c_i e^{\lambda_i t}\mathbf{r}_i, \tag{5.129}$$

where $c_i = \text{const}$, $\mathbf{r}_i$ is the eigenvector of J, and λ_i is the corresponding eigenvalue. Consider the *ODE system with decay*: $\text{Re}(\lambda_i) < 0$, with the corresponding timescales $1/\text{Re}(-\lambda_i)$. The system with a decay is called *stiff* [157], if all timescales are considerably different:

$$\frac{\max_i \text{Re}(-\lambda_i)}{\min_i \text{Re}(-\lambda_i)} \gg 1. \tag{5.130}$$

Such stiff system describes an interesting class of physical problems with decays and different timescales. For example, the reaction rates of elementary particles in the *electron–positron plasma* have very different timescales, see Chapter 14.

In order to insure the stability, one has to use such time step that for all eigenvalues $\Delta t\lambda_i$ is in the region of stability. For the methods with the finite region of stability, the time step is limited by the smallest timescale of the system, while the total time of the integration is similar to the maximal timescale. This implies that *explicit methods* are not adequate here. To avoid limits on the time step, the *A-stable methods* are useful. The method is called A-stable if its region of stability includes all the half-plane $\text{Re}(\Delta t\lambda) < 0$ [157].

The linear multistep *implicit methods* can be stable only for the order less or equal to 2 [157]. An example of the second-order *implicit method* is the *trapezoidal rule*:

$$\frac{y^{n+1} - y^n}{\Delta t} = \frac{f(y^n, t^n) + f(y^{n+1}, t^{n+1})}{2}. \tag{5.131}$$

For the equation $\dot{y} = \lambda y$, the *trapezoidal rule* gives $y^{n+1} = \frac{1+\lambda\Delta t/2}{1-\lambda\Delta t/2}y^n$. The region of the stability is

$$\left|1 + \frac{\Delta t\lambda}{2}\right| \leq \left|1 - \frac{\Delta t\lambda}{2}\right|, \tag{5.132}$$

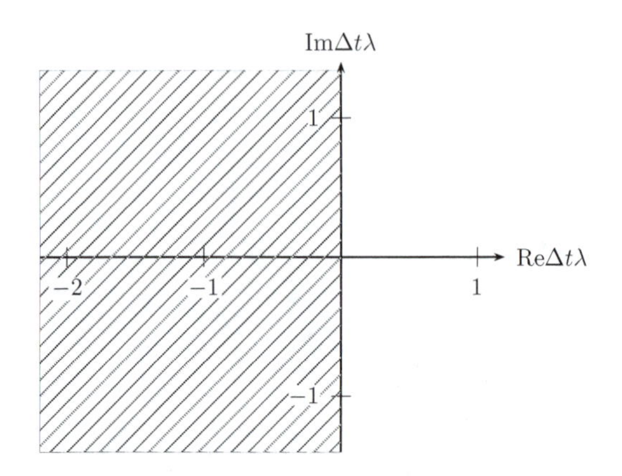

Figure 5.12 Implicit *trapezoidal rule* is stable in the filled region.

or when $\Delta t\lambda/2$ is closer to -1 than to 1, which is the entire left complex half-plane $\mathrm{Im}\Delta t\lambda \leq 0$ (Fig. 5.12). Another *A-stable method* is implicit Euler one

$$\frac{y^{n+1}-y^n}{\Delta t} = f(y^{n+1}, t^{n+1}), \tag{5.133}$$

with the stability condition

$$\left|1 - \frac{\Delta t\lambda}{2}\right|^{-1} \leq 1. \tag{5.134}$$

Its stability region is the exterior of the circle of radius 1 centered at $(1,0)$ in the complex plane $\Delta t\lambda$, see Fig. 5.13.

A-stability is a very strong condition, and in practice less strong conditions are adopted. The method is called $A(a)$ stable with $a \in (0, \pi/2)$ if the region of the stability is $|\arg(-\Delta t\lambda)| < a$, see Fig. 5.14. Gear [137] introduced the definition of *stiff stability*:

- the method is absolutely stable in the region R_1 $(\mathrm{Re}(\Delta t\lambda) \leq D)$ and
- the method is exact in the region R_2 $(D < \mathrm{Re}(\Delta t\lambda) < a,$ $|\mathrm{Im}(\Delta t\lambda)| < \theta)$.

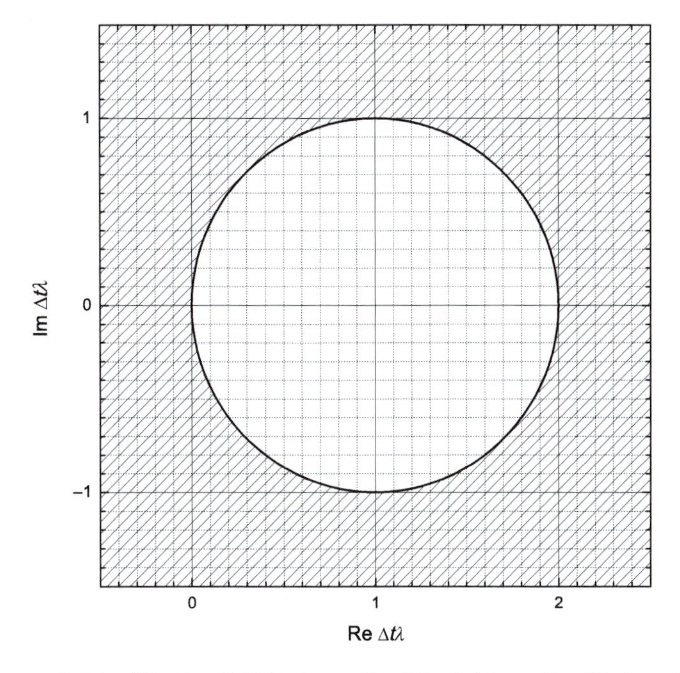

Figure 5.13 The implicit Euler method is stable in the filled region.

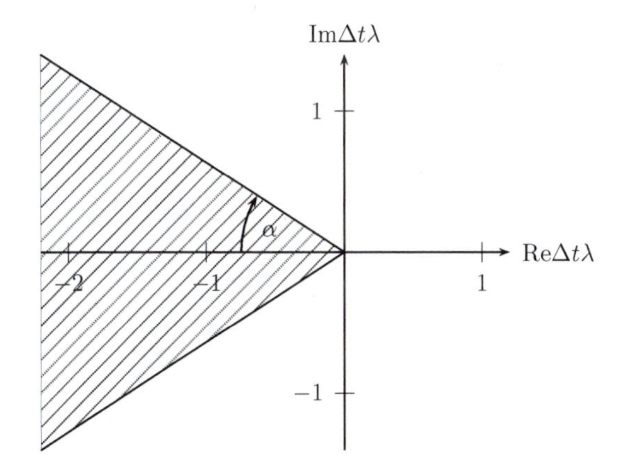

Figure 5.14 $A(\alpha)$ stable method in the filled region.

Gear's method is the implicit *predictor–corrector method* with backward differences:

$$y^{n+1} = \sum_{j=1}^{k} \alpha_j y^{n-j+1} + \Delta t \beta_0 f(t^{n+1}, y^{n+1}), \tag{5.135}$$

coefficients of the method α_j, β can be found in Ref. [137] for the different orders of the method k, $1 \leq k \leq 6$. The difference (5.135) gives the system of nonlinear equations for y^{n+1} to be solved with *Newton's iterations.* It is possible to variate the order k from 1 to 6 to achieve the maximal time step for the given accuracy of the integration. The method also utilizes *Jacobi matrix* from the previous time steps when possible in order to reduce number of operations for the *matrix inversions.* The regions of stability of *Gear's method* [137] are illustrated in Fig. 5.15. The curve

$$\Delta t \lambda(\theta) = -\frac{-e^{iq\theta} + \sum_{j=1}^{q} \alpha_j e^{i(q-j)\theta}}{\beta_0 e^{iq\theta}}, \quad 0 \leq \theta \leq 2\pi, \tag{5.136}$$

is the boundary of the region of the stability in Fig. 5.15.

5.7 Numerical methods for linear algebra

Once the numerical solution of the system of PDEs is reduced to the system of ODEs, the methods of linear algebra can be utilized. In this section, a brief discussion of such methods is given.

5.7.1 *Exact solution*

Consider a system of linear algebraic equations

$$A\mathbf{u} = \mathbf{F}, \tag{5.137}$$

where A is $n \times n$ *square matrix.* It can be solved by the *Gauss elimination* method, which is the exact method to find the inverse matrix A^{-1} and the solution for vector $\mathbf{u}$. First, the $n \times n$ identity matrix $E = \mathrm{diag}\{1, \ldots, 1\}$ is augmented to the right of A, forming a $n \times 2n$ block matrix $(A|E)$. Be means of elementary transformations, the composite matrix $(A|E)$ is reduced to the form $(E|A^{-1})$, where A^{-1} is the inverse matrix. The elementary transformations are the linear combinations of the matrix rows. The unknown vector $\mathbf{u}$ can then be found from the relation $\mathbf{u} = A^{-1}\mathbf{F}$. The number of operations is $\propto N^2$.

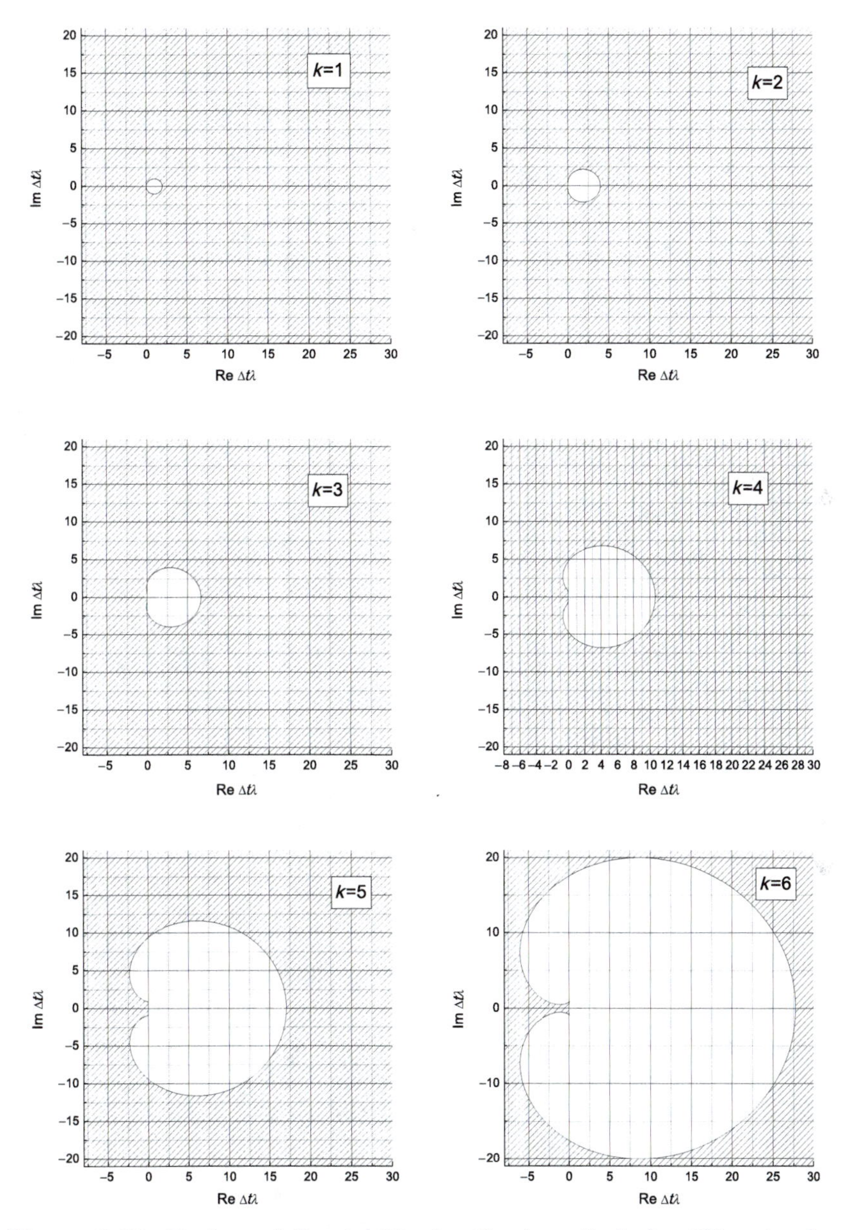

Figure 5.15 Regions of the stability for *Gear's methods* for different orders $k = 1, 2, 3, 4, 5, 6$.

Often at the three-point approximation of space derivatives in *implicit schemes*, one deals with the system of linear equation with tridiagonal-type matrix A

$$a_i u_{i-1} + b_i u_i + c_i u_{i+1} = f_i, \quad 1 \leq i \leq N \tag{5.138}$$

for the unknown vector u_i. For such a system, the *cyclic reduction method* is useful [143,260]. In this method, one searches the relations between unknown components of the solution in the linear form:

$$u_{i+1} = x_{i+1} u_i + y_{i+1}. \tag{5.139}$$

It is possible to derive coefficients x_i, y_i from coefficients x_{i+1} and $y_{i=1}$ by inserting the solution (5.139) in the initial Eq. (5.138):

$$a_i u_{i-1} + b_i u_i + c_i(x_{i+1} u_i + y_{i+1}) = f_i. \quad 1 \leq i \leq N. \tag{5.140}$$

One obtains

$$u_i = -\frac{a_i}{b_i + c_i x_i} u_{i-1} + \frac{f_i - c_i y_{i+1}}{b_i + c_i x_{i+1}} = x_i u_{i-1} + y_i, \tag{5.141}$$

and

$$x_i = -\frac{a_i}{b_i + c_i x_{i+1}}, \qquad y_i = \frac{f_i - c_i y_{i+1}}{b_i + c_i x_{i+1}}. \tag{5.142}$$

The right boundary condition

$$a_N u_{N-1} + b_N u_N = f_N \tag{5.143}$$

allows one to calculate coefficients x_N, y_N. The number of operations is proportional to the rank of matrix $\propto N$. The generalization to block tridiagonal matrix A is straightforward. The *cyclic reduction method* is used in the calculations in Part 3. It is possible to realize a parallel algorithm for the tridiagonal matrix cyclic solver [162], see also Ref. [5].

5.7.2 *Iterative methods*

In most cases, one needs to solve algebraic equations with a large number of unknowns, as large as the number of volumes on the *computational grid*. The iterative methods can be useful in some cases.

5.7.2.1 *Jacobi method*

The *Jacobi method* is a simple algorithm for determination of the solutions of a diagonally dominant system of linear equations [143]. The initial matrix A is represented by the sum of two matrices: the diagonal matrix D and the matrix R with zero elements on the diagonal

$$A = D + R, \text{ where } D = \begin{pmatrix} a_{11} & 0 & \dots & 0 \\ 0 & a_{22} & \dots & 0 \\ \vdots & \vdots & \ddots & \vdots \\ 0 & 0 & \dots & a_{nn} \end{pmatrix},$$

$$R = \begin{pmatrix} 0 & a_{12} & \dots & a_{1n} \\ a_{21} & 0 & \dots & a_{2n} \\ \vdots & \vdots & \ddots & \vdots \\ a_{n1} & a_{n2} & \dots & 0 \end{pmatrix}. \tag{5.144}$$

The iteration procedure for the new $k+1$th iteration is

$$\mathbf{u}^{(k+1)} = D^{-1}(\mathbf{x} - R\mathbf{x}^{(k)}),$$

or for the elements,

$$u_i^{(k+1)} = \frac{1}{a_{ii}}\left(b_i - \sum j \neq i a_{ij} u_j^k\right).$$

The diagonal dominance means that for each row, the absolute value of the diagonal term is greater than the sum of absolute values of other terms:

$$|a_{ii} > \sum_{j \neq i} |a_{ij}|.$$

5.7.2.2 *Gauss–Seidel method*

The idea of the iterative *Gauss–Seidel method* is connected with the simplicity of the inverse matrix for the triangular matrix.

For the equation

$$A\mathbf{u} = \mathbf{f}, \tag{5.145}$$

one can write the iteration process

$$L_* \mathbf{u}^{(k+1)} = \mathbf{f} - U\mathbf{u}^{(k)}. \tag{5.146}$$

The decomposition of A into the lower triangle L_* and strictly upper triangle U matrices is performed:

$$A = L_* + U, \quad \text{where } L_* = \begin{pmatrix} a_{11} & 0 & \dots & 0 \\ a_{21} & a_{22} & \dots & 0 \\ \vdots & \vdots & \ddots & \vdots \\ a_{n1} & a_{n2} & \dots & a_{nn} \end{pmatrix},$$

$$U = \begin{pmatrix} 0 & a_{12} & \dots & a_{1n} \\ 0 & 0 & \dots & a_{2n} \\ \vdots & \vdots & \ddots & \vdots \\ 0 & 0 & \dots & 0 \end{pmatrix}. \tag{5.147}$$

The new $(k+1)$ iteration in components is

$$u_i^{(k+1)} = \frac{1}{a_{ii}} \left(f_i - \sum_{j<i} a_{ij} x_j^{(k+1)} - \sum_{j>i} a_{ij} u_j^{(k)} \right). \tag{5.148}$$

The procedure is known to converge [143] if either A is symmetric and positive-definite

$$\mathbf{z}^T A \mathbf{z} > 0 \tag{5.149}$$

for non-zero vector $\mathbf{z}$ or A is strictly or irreducibly diagonally dominant

$$|a_{ii}| \geq \sum_{j \neq i} |a_{ij}|. \tag{5.150}$$

5.7.2.3 *Successive over-relaxation method*

In the *successive over-relaxation method*, one adopts the decomposition of A into a diagonal D, and strictly lower and upper triangle components L, U:

$$A = D + L + U, \tag{5.151}$$

where

$$D = \begin{pmatrix} a_{11} & 0 & \dots & 0 \\ 0 & a_{22} & \dots & 0 \\ \vdots & \vdots & \ddots & \vdots \\ 0 & 0 & \dots & a_{nn} \end{pmatrix}, \quad L = \begin{pmatrix} 0 & 0 & \dots & 0 \\ a_{21} & 0 & \dots & 0 \\ \vdots & \vdots & \ddots & \vdots \\ a_{n1} & a_{n2} & \dots & 0 \end{pmatrix},$$

$$U = \begin{pmatrix} 0 & a_{12} & \dots & a_{1n} \\ 0 & 0 & \dots & a_{2n} \\ \vdots & \vdots & \ddots & \vdots \\ 0 & 0 & \dots & 0 \end{pmatrix}. \tag{5.152}$$

The system of linear equations may be rewritten as follows:

$$(D + \omega L)\mathbf{u} = \omega\mathbf{f} - [\omega U + (\omega - 1)D]\mathbf{u} \tag{5.153}$$

with the *relaxation factor* $\omega > 1$. The new iteration is

$$\mathbf{u}^{(k+1)} = (D+\omega L)^{-1}[\omega\mathbf{f} - (\omega U + (\omega - 1)D)\mathbf{u}^{(k)}] = L_\omega \mathbf{x}^{(u)} + \mathbf{c}, \tag{5.154}$$

or for elements

$$u_i^{(k+1)} = (1-\omega)u_i^{(k)} + \frac{\omega}{a_{ii}}\left(b_i - \sum_{j<i} a_{ij}u_j^{(k+1)} - \sum_{j>i} a_{ij}u_j^{(k)}\right),$$

$$i = 1, 2, \dots, n. \tag{5.155}$$

If A is symmetric and positive-definite, then $\rho(L_\omega) < 1$ for $0 < \omega < 2$ and the convergence of the iteration follows [130,143]. In general, for $\omega = 1$, one should choose the Seidel method, for $\omega > 1$, the successive over-relaxation one, and for $\omega < 1$, the successive down-relaxation one.

5.7.2.4 *Iterative methods for the sparse matrix*

For the computation of a number of problems, one needs the inverse of a sparse matrix. A *sparse matrix* is a matrix in which most of the elements are zero. In contrast, when most of the elements are non-zero, the matrix is said to be dense. The fraction of non-zero elements over the total number of elements (i.e., that can fit into the matrix, say a matrix of dimension of $m \times n$ can accommodate $m \times n$ total number of elements) in a matrix is called the sparsity (density). In some cases, the structure of the matrix is well known. For example, in the direct integration of the kinetic Boltzmann equation by the implicit code, one has block matrices like the three-diagonal matrices. In that case, the *cyclic reduction method* is effective.

Numerous problems operate with the *sparse matrix* with the common structure. For the example, in Ref. [12], a *Newton's iteration* method for obtaining equilibria of rapidly rotating axially symmetric stars was developed in the base sparse matrix solver. The Bernoulli integral for the barotropic *equation of state* and the *Poisson equation* for the gravitational potential in the *computational grid* were reduced to the set of *finite difference equation* for the dimensionless gravitational potential. The system of nonlinear equations for the gravitational potential was solved by *Newton's iteration* method. The system of linear algebra equations at every step of the iteration was solved by the elimination technique developed by Zlatev [248]. This approach allows one to operate with the huge sparse matrices and uses *LU* decomposition iteration process.[5] Such common solver for the *sparse matrix* is less effective in comparison with the matrix elimination of the sparse matrix with the known structure.

[5]See also the realization the NAG Library on the website: https://nag.com/nag-library/.

Chapter 6

Direct Integration of Boltzmann Equations

In this chapter, our experience of the integration of Boltzmann equations by finite difference method is presented. In spherically symmetric geometry, the approach operates with finite differences on the fixed grid in 4D *phase space* for particles (r, μ, ϵ, t), and it is based on the *method of lines*. The ODE system is solved by implicit *Gear's method* suitable for integration of *stiff equations*. The method takes into account all reactions and is applicable to both optically thick and optically thin regions.

6.1 Finite differences and the method of lines

This section is limited to 1D spatial geometry (spherically symmetric case) with full physical processes and *implicit methods*.

A large number of physical problems requires the solution of Boltzmann equations. Analytical solutions are available in exceptional cases, and in general case, one should rely on effective numerical integration. As discussed in the previous chapter, finite difference technique [284] represents one of such effective methods.

Finite differences are widely employed in astrophysical problems. Specific examples, considered in Part 3, include thermalization of non-equilibrium optically thick *electron–positron plasma* and neutrino transport in the core collapse *supernovae*.

Finite differences for the kinetic equation in astrophysical applications were introduced in Ref. [220]. In particular, in the *gravitational collapse* of the iron core of a massive star to a *neutron star*, neutrinos play a crucial role. Neutrino transport should be described by Boltzmann equations. The core on the verge of collapse has the radius 10^8 cm and it is transparent for neutrinos. The collapsed hot protoneutron star has a radius of 10^6–10^7 cm and neutrinos are trapped with the *optical depth* $\gtrsim 10^3$. It is impossible to treat such problems by *explicit methods* due to the presence of different timescales characterizing different processes.

The relativistic Boltzmann equation (2.18) for the distribution function in the spherically symmetric flow $f_\iota(|\mathbf{p}|, \mu, r, t)$ for the particle of sort ι, see, e.g., Refs. [118,220,224], is

$$\frac{1}{c}\frac{\partial f_\iota}{\partial r} + \beta_\iota\left(\mu\frac{\partial f_\iota}{\partial r}\frac{1-\mu^2}{r}\frac{\partial f_\iota}{\partial \mu}\right) = \sum_q (\tilde{\eta}_\iota^q - \chi_\iota^q f_\iota), \tag{6.1}$$

where $\mu = \cos\vartheta$, ϑ is the angle between the radius vector $\mathbf{r}$ and the particle momentum $\mathbf{p}$. In addition, $\tilde{\eta}_\iota^q$ is the *emission coefficient* for the production of a particle of sort ι via the physical process labeled by q, and χ_ι^q is the corresponding *absorption coefficient*. The summation runs over all considered physical processes that involve a particle of sort ι. Gravity and external forces are neglected in (6.1). Particle densities are determined from Eq. (2.5).

In numerical simulations, it is convenient to represent hyperbolic PDEs in the so-called *conservative form*. The *"conservative" numerical method* can provide exact conservation of energy on a finite computational grid. The Boltzmann equations are said to be in conservative form when the advection terms on the LHS are represented as a derivative of a flux divided by a volume [159]. In order to take this form, the quantities

$$E_\iota(\epsilon_\iota, \mu, r, t) = \frac{2\pi\epsilon_\iota^3\beta_\iota f_\iota}{c^3}, \tag{6.2}$$

are used instead of f_ι. Since

$$\epsilon_\iota f_\iota d\mathbf{r}d\mathbf{p} = \frac{2\pi\epsilon_\iota^3\beta_\iota f_\iota}{c^3} d\mathbf{r}d\epsilon_\iota d\mu = E_\iota d\mathbf{r}d\epsilon_\iota d\mu, \tag{6.3}$$

it is clear that E_ι is the *spectral energy density* in the $\{\mathbf{r}, \mu, \epsilon_\iota\}$ *phase space*, in which the volume element is $d\mathbf{r}d\epsilon_\iota d\mu$. Using

Eqs. (6.1) and (6.3), the Boltzmann equations can be written in terms of the *spectral energy density* E_ι as

$$\frac{1}{c}\frac{\partial E_\iota}{\partial r}+\frac{\mu}{r}\frac{\partial}{\partial r}(r^2\beta_\iota E_\iota)+\frac{1}{r}\frac{\partial}{\partial \mu}\left[(1-\mu^2)\beta_\iota E_\iota\right]=\sum_q(\eta_\iota^q-\chi_\iota^q E_\iota), \quad (6.4)$$

where $\eta_\iota^q = 2\pi\epsilon_\iota^3\beta_\iota\tilde{\eta}_\iota^q/c^3$. This form of Boltzmann equation is the basis for the conservative finite difference code used for numerical simulations reported in Part 3. It allows one to carry out calculations with large time steps even for large *optical depths.*

A *computational grid* in the $\{\mathbf{r},\mu,\epsilon\}$ *phase space* is defined as follows. The r domain ($R < r < r_{\rm ext}$) is divided into $j_{\rm max}$ spherical shells whose boundaries are designated with half integer indices. The jth shell ($1 \leq j \leq j_{\rm max}$) is between $r_{j-1/2}$ and $r_{j+1/2}$, with $\Delta r_j = r_{j+1/2}-r_{j-1/2}$. The μ-grid is made of $k_{\rm max}$ intervals $\Delta\mu_k = \mu_{k+1/2}-\mu_{k-1/2}$: $1 \leq k \leq k_{\rm max}$. The energy grids for different kinds of particles are different. The quantities to be computed are the energy densities averaged over phase-space cells

$$E_{\iota,\omega,k,j}(t)=\frac{1}{\Delta X}\int_{\Delta\epsilon_\omega,\Delta\mu_k,\Delta r_j}E_\iota d\epsilon d\mu r^2 dr, \quad (6.5)$$

where $\Delta X = \Delta\epsilon_\omega\Delta\mu_k\Delta(r_j^3)/3$ and $\Delta(r_j^3) = r_{j+1/2}^3 - r_{j-1/2}^3$.

Replacing the space and angle derivatives in the Boltzmann equation (6.4) by *finite differences*, one arrives at the following set of ODEs for the quantities $E_{i,\omega,k,j}$ specified on the *computational grid*:

$$\frac{dE_{\iota,\omega,k,j}}{dt}+\beta_{\iota,\omega}\frac{\Delta(r^2\mu_k E_{\iota,\omega,k})_j}{\Delta(r_j^3)/3}+\left\langle\frac{1}{r}\right\rangle_j\beta_{\iota,\omega}\frac{\Delta\left[(1-\mu^2)E_{\iota,\omega,j}\right]_k}{\Delta\mu_k}$$
$$=\sum_q[\eta_{\iota,\omega,k,j}^q-(\chi E)_{\iota,\omega,k,j}^q], \quad (6.6)$$

where $\beta_{\gamma,\omega}=1$ and $\beta_{e,\omega}=\left[1-(m_e c^2/\epsilon_{\iota,\omega})^2\right]^{1/2}$. In order to achieve the *second-order approximation*, the following quantities are replaced by their mean values over the volume or over the corresponding coordinate on the grid or the *central differences*:

$$\epsilon_{\iota,\omega} = \frac{\epsilon_{\iota,\omega-1/2} + \epsilon_{\iota,\omega+1/2}}{2}, \tag{6.7}$$

$$\mu_k = \frac{\mu_{k-1/2} + \mu_{k+1/2}}{2}, \tag{6.8}$$

$$\left\langle \frac{1}{r} \right\rangle_j = \frac{(r_{j+1/2}^2 - r_{j-1/2}^2)/2}{(r_{j+1/2}^3 - r_{j-1/2}^3)/3}, \tag{6.9}$$

$$r_j = \frac{r_{1-1/2} + r_{j+1/2}}{2}, \tag{6.10}$$

$$E_{\iota,\omega,k}(r) = \frac{1}{\Delta\epsilon_\omega \Delta\mu_k} \int_{\Delta\epsilon_\omega \Delta\mu_k} E_\iota(\epsilon, \mu, r) d\epsilon d\mu, \tag{6.11}$$

$$E_{\iota,\omega,j}(\mu) = \frac{3}{\Delta\epsilon_\omega \Delta r_j^3} \int_{\Delta\epsilon_\omega \Delta r_j} E_\iota(\epsilon, \mu, r) dr d\epsilon, \tag{6.12}$$

$$E_{\iota,k,j}(\epsilon) = \frac{3}{\Delta\mu_k \Delta r_j^3} \int_{\Delta\mu_k \Delta r_j^3} E_\iota(\epsilon, \mu, r) dr d\mu, \tag{6.13}$$

$$\begin{aligned} \Delta(r^2 \mu_k E_{\iota,\omega,k})_j &= r_{j+1/2}^2 (\mu_k E_{\iota,\omega,k})_{r=r_{j+1/2}} \\ &\quad - r_{j-1/2}^2 (\mu_k E_{\iota,\omega,k})_{r=r_{j-1/2}}, \end{aligned} \tag{6.14}$$

$$\begin{aligned} \Delta\left(\epsilon\beta_\iota E_{\iota,k,j}\right)_\omega &= \epsilon_{\iota,\omega+1/2}\beta_{\iota,\omega+1/2}(E_{\iota,k,j})_{\omega+1/2} \\ &\quad - \epsilon_{\iota,\omega-1/2}\beta_{\iota,\omega-1/2}(E_{\iota,k,j})_{\omega-1/2}, \end{aligned} \tag{6.15}$$

$$\begin{aligned} \Delta\left[(1-\mu^2)E_{\iota,\omega,j}\right]_k &= (1-\mu_{k+1/2}^2)(E_{\iota,\omega,j})_{\mu=\mu_{k+1/2}} \\ &\quad -(1-\mu_{k-1/2}^2)(E_{\iota,\omega,j})_{\mu=\mu_{k-1/2}}. \end{aligned} \tag{6.16}$$

The LHS of the Boltzmann equation in the quasi-linear form is the PDE of the hyperbolic type. For numerical solution, one can use the *"upwind" derivative approximation*[1] for derivatives along μ and ϵ

[1]This is integration along *characteristics* in 1D case.

without introducing the *artificial viscosity*

$$(E_{\iota,k,j})_{\omega+1/2} = \begin{cases} E_{\iota,\omega+1,k,j} + \dfrac{(E_{\iota,\omega+2,k,j} - E_{\iota,\omega+1,k,j})(\epsilon_{\omega+1/2} - \epsilon_{\omega+1})}{\epsilon_{\omega+2} - \epsilon_{\omega+1}}, \\ \qquad \mu \geq 0 \\ E_{\iota,\omega,k,j} + \dfrac{(E_{\iota,\omega,k,j} - E_{i,\omega-1,k,j})(\epsilon_{\omega+1/2} - \epsilon_{\omega})}{\epsilon_{\omega} - \epsilon_{\omega-1}}, \\ \qquad \mu < 0, \end{cases} \tag{6.17}$$

$$\left(E_{\iota,\omega,j}\right)_{\mu=\mu_{k+1/2}} = E_{\iota,\omega,k,j} + \frac{\Delta\mu_k(E_{\iota,\omega,k,j} - E_{\iota,\omega,k-1,j})}{\Delta\mu_{k-1} + \Delta\mu_k}, \tag{6.18}$$

which should be restricted to non-negative values. For the approximation of the derivative over r, the second-order "upwind" differences cannot be used. This is because the method employs the *cyclic reduction* for the solution of the ODE system. Instead, the combination of the *first-order approximation "upwind" difference* and the *second-order central difference* is adopted

$$\begin{aligned}(\mu_k E_{\iota,\omega,k})_{r=r_{j+1/2}} \\ &= (1 - \tilde{\chi}_{\iota,\omega,k,j+1/2})\left(\frac{\mu_k + |\mu_k|}{2} E_{\iota,\omega,k,j} + \frac{\mu_k - |\mu_k|}{2} E_{\iota,\omega,k,j+1}\right) \\ &\quad + \tilde{\chi}_{\iota,\omega,k,j+1/2}\mu_k \frac{E_{\iota,\omega,k,j} + E_{\iota,\omega,k,j+1}}{2}\end{aligned} \tag{6.19}$$

with the coefficient

$$\tilde{\chi}^{-1}_{\iota,\omega,k,j+1/2} = 1 + \frac{1}{\chi_{\iota,\omega,k,j}\Delta r_j} + \frac{1}{\chi_{\iota,\omega,k,j}\Delta r_{j+1}}. \tag{6.20}$$

The dimensionless coefficient $\tilde{\chi}$ is introduced to describe correctly both the optically thin and the optically thick computational cells by means of a compromise between the high-order method and the monotonic transport scheme without the *artificial viscosity*, see Refs. [7,220,267]. In the *optically thin region*, the scheme is first-order "upwind". In the *optically thick region*, the scheme different. Numerical oscillations characteristic for the non-monotonic scheme are suppressed by high reaction rates, which on the grid become

$$\eta^q_{\iota,\omega,k,j} = \frac{1}{\Delta X}\int_{\Delta\epsilon_\omega,\Delta\mu_k,\Delta r_j} \eta^q_\iota d\epsilon d\mu r^2 dr, \tag{6.21}$$

$$(\chi E)^q_{\iota,\omega,k,j} = \frac{1}{\Delta X}\int_{\Delta\epsilon_\omega,\Delta\mu_k,\Delta r_j} \chi_\iota E^q_\iota d\epsilon d\mu r^2 dr. \tag{6.22}$$

The physical processes included in the simulations and the expressions $\eta^q_{\iota,\omega,k,j}$ and $(\chi E)^q_{\iota,\omega,k,j}$ are discussed in Part 3 and the Appendices. Clearly, in the numerical scheme, $\eta^q_{\iota,\omega,k,j}$ and $(\chi E)^q_{\iota,\omega,k,j}$ are sums instead of integrals of the Boltzmann equations.

The numerical method of the conversion of the evolutionary PDEs into the ODEs is the so-called *method of lines* [73]. The method of the numerical solution of the ODEs can be either explicit or implicit. When there are different timescales in the physical problem, one should use *implicit schemes* such as *Gear's method* suitable for the numerical integration of *stiff ODE systems*. For the inverse of the *sparse matrix* with the structure $I + \Delta t J$, where J is the *Jacobi matrix*, the *cyclic reduction method* is useful for the three-diagonal matrix with rank $j_{\max}$ composed from the matrices $k_{\max}\times\omega_{\max}\times\iota_{\max}$. Note that the expansion of the presented method to 2D geometry in coordinate space is an open question, due to limitations related to the use of *cyclic reduction method* with three-diagonal matrices for solution of the system of linear algebraic equations.

Chapter 7

Multidimensional Hydrodynamics

This chapter describes multidimensional hydrodynamics for multi-temperature mixture of gases. The classical multidimensional shock capturing hydrodynamics and the application of modern high-order Godunov-type methods reduce the problem into a class of almost engineering tasks.

Scientific input includes *equation of state*, reaction rates, and kinetic coefficients, described in the second section. The main difficulties of kinetic Boltzmann approach discussed in the previous chapter are not only the multidimensionality of the *phase space* but also the calculation of the reaction rates. These reaction rates usually require to use *implicit schemes* in the case of non-transparent regions, and multidimensional problems become very hard to study. The key point of this section is the proposal to move from the kinetic Boltzmann treatment in 7D *phase space* $(\mathbf{r}, \mathbf{p}, t)$ to hydrodynamic one with diffusion and flux limiters in 5D *phase space* $(\mathbf{r}, \epsilon, t)$. The *diffusion with flux limiter* approach uses some free parameters for the interpolation of spectral energy fluxes in the intermediate case between the transparent (free flow) and the non-transparent (diffusion or heat conduction) regions. The first calculations within such approach were performed in Ref. [84] for the *gravitational collapse* with the neutrino transport in spherically symmetric case. In such a problem, one has to carry out advection by explicit code. At the same time, diffusion of spectral energies is carried out by *implicit scheme*, such as *Crank–Nicolson method* [116]. Finally, reaction rates are computed using *implicit scheme* for the system of ODEs. In the multidimensional 2D or 3D cases, the splitting on the time integration along separate directions is made. Given the limited space resolution in 3D case in comparison with 1D one, high-order Godunov-type methods are required.

The common idea in this chapter is the multidimensional hydrodynamics and *explicit methods* for advection.

7.1 High-order Godunov methods

In the previous chapter, the universal implicit scheme for numerical solution of Boltzmann equations suitable for spherically symmetric geometry was presented. In 3D space, the distribution functions are defined on 7D *phase space* $f_\iota(\mathbf{r}, \mathbf{p}, t)$. From the computational point of view, it is not possible to realize the implicit finite difference method for the evaluation of distribution functions $f_\iota(\mathbf{r}, \mathbf{p}, t)$ in such multidimensional case. Significant simplification of the problem is required.

First calculations of the *gravitational collapse* used neutrino heat conductivity in the non-transparent region with a smooth transition to the transparent regions in 1D [175]. *Diffusion with flux limiter* for the calculations of the hydrodynamic equations in the continuous computational region was adopted in Refs. [84,122]. In particular, the idea of *"flux limiter multigroup diffusion"* for the spectral neutrino energy densities $\rho\epsilon_\iota$ transport, instead of the description for distribution functions $f_\iota(r, \mu, \epsilon, t)$, was applied. The mathematical problem includes the hydrodynamical equations: the mass conservation, the momentum conservation, the energy conservation, additional *advection equations* for the neutrino energy, energy diffusion, and reaction rates on the right-hand side of equations. Such approach with the diffusion of energy of neutrino is adequate in the nontransparent region. In the transparent regions, it is possible to introduce flux limiters to obtain the energy flux, corresponding to the flow in the vacuum. For the whole computational region, the smooth interpolation of the energy fluxes between transparent and non-transparent regions is used in Ref. [84].

Such simplified approach is applicable for the multidimensional gravitational core collapse with the neutrino transport and in the simulation of micro targets in the inertial fusion with the important role of the energy transfer by photons and other "fast" particles. The time integration in multidimensional case is performed by the splitting on the time integration along separate directions. The hydrodynamic advection can be carried out within the framework of the explicit hydrodynamic methods with the limit on time step given by the *Courant condition* [114] $\sim\Delta r/(c_s + v)$, where Δr is the grid size, c_s is the *sound speed*, and v is the fluid velocity. The energy diffusion and the energy exchange can be considered in the framework of the *method of lines* and the *implicit scheme* for the integration of the ODE system. Formally, the *implicit scheme* does not limit time steps, but in the case of transparent region, the accuracy of the integration gives time limit $\sim\Delta r/c$. Nonlinear diffusion in the *optically thin region* reduces to the energy transfer with the speed of light c.

In modern hydrodynamic simulations with the application in astrophysics, the *Godunov high-order methods* are the most effective ones. The reason is simple. In 1D space, it is possible to use the *Lagrangian approach* based on the *comoving coordinates* (m, t). In the multidimensional case with instabilities, initial orthogonal *computational grid* deforms in the course of evolution and the spatial accuracy gets lost. In 2D and 3D cases, the Eulerian fixed computational grid is preferred.

In astrophysical simulations, the physical quantities such as density and *pressure* can vary by several orders of magnitude. In such applications, *Godunov high-order methods* have important advantages: The method provides the accurate representation of *contact discontinuities*; the method can automatically change the space order of the approximation and provides the *monotonic solution* without the false oscillations (giving unphysical negative density or specific energy) in the region with high spatial gradients. In the following, the details of Godunov methods in 1D problems are discussed. The time integration in multidimensional case is obtained by the *dimensional splitting* method.

In 1957, Godunov [141] proposed a shock capturing method applicable to gas-dynamic simulation with discontinuities and SWs. The method is based on the introduction of a spatial grid in the computational domain on which volume-averaged independent physical quantities $\mathbf{U}(\mathbf{r}, t)$ are defined at the time step $t = t^n$. The gas-dynamic equations are written in the form of *conservation laws* for conservative quantities (see Appendix A):

$$\frac{\partial \mathbf{U}}{\partial t} + \mathrm{div}\mathbf{F}(\mathbf{U}) = 0. \tag{7.1}$$

The differential equations are regarded as *conservation laws* since the nonlinear gas-dynamic equations can lead to discontinuous solutions even in the case of smooth initial conditions [269]. In terms of finite differences, this means that arbitrary volumes ΔV_j satisfy the equation

$$\frac{\partial \mathbf{U}_j}{\partial t} + \sum_k \frac{\mathbf{F}_{jk} \Delta \mathbf{s}_{jk}}{\Delta V_j} = 0, \tag{7.2}$$

where the sum of the fluxes $\mathbf{F}_{jk}$ is taken over all surfaces $\Delta \mathbf{s}_{jk}$ around the volume ΔV_j. After introducing the time step $\Delta t = t^{n+1} - t^n$, one can write down the explicit *finite difference scheme*

$$\frac{\mathbf{U}_j^{n+1} - \mathbf{U}_j^n}{\Delta t} + \sum_k \frac{\langle \mathbf{F}_{jk} \rangle \Delta \mathbf{s}_{jk}}{\Delta V_j} = 0, \tag{7.3}$$

where $\langle \mathbf{F}_{jk} \rangle$ denotes the time-averaged fluxes through the cell boundaries. Actually, Godunov considered a 1D problem in Lagrangian variables (m, t). However, for pedagogical purposes, it is better to state the problem in multidimensional fixed Euler coordinates $(\mathbf{r}, t)$.

Godunov's idea was to obtain time-averaged fluxes by solving the *Riemann problem.* This is the 1D problem of finding a self-similar solution (depending on the parameter x/t) given two constant gas states to the left and to the right of the coordinate $x = 0$. For the *explicit scheme* to be stable, it must satisfy the *Courant condition*, i.e., the time step is bounded and the region of influence has to be taken into account. The scheme has first *order of accuracy* if, as in the original version, $\mathbf{U}(\mathbf{r}, t) = \mathbf{U}_j^n$ are assumed to be constant inside the cells over the entire integration step. Godunov conducted research and proved that, among linear schemes, only *first-order approximations* are monotonicity preserving, i.e., at every step, a *monotonic solution* remains so without generating spurious oscillations (see Ref. [142]). The scheme did not require the explicit introduction of numerical viscosity. The *Riemann problem* was considered for an ideal gas.

Later, the order of Godunov-type schemes was increased due to the reconstruction of the solution $\mathbf{U}(\mathbf{r}, t)$ inside cells with the use of the data from the neighboring cells $\mathbf{U}_j^n$, see, e.g., Ref. [108]. Monotonicity was ensured by requiring that there be no local minima or maxima in the interpolation, and the time-averaged fluxes $\langle \mathbf{F}_{jk} \rangle$ were determined by solving the *Riemann problem* for prepared averaged solutions from the Riemann invariants to the left and to the right of the boundaries, i.e., the high order for small perturbations in Godunov schemes was combined with monotonicity. In numerical experiments, this means that SW of arbitrary intensity can be computed without going beyond the physically admissible limits of grid functions and that *contact discontinuities* are only slightly smeared on the computational grid.

The ideal gas equation of state was used in the original scheme. The method was extended [107,268] to an arbitrary *tabulated equation of state* $P = P(\rho, \epsilon)$, $\epsilon = \epsilon(\rho, T)$. Specifically, a *linearized Riemann problem* was considered in Ref. [268], while an efficient approximate Riemann solver based on a *local model for the equation of state* was constructed in Ref. [107].

7.2 Multidimensional multitemperature high-order Godunov code

An issue of special interest in physics is concerned with problems for a *multicomponent gas* of different substances ι described by a set of densities $\rho_\iota(\mathbf{r}, t) \equiv c_\iota(\mathbf{r}, t)\rho(\mathbf{r}, t)$, where c_ι are concentrations, $\rho\epsilon_\iota(\mathbf{r}, t)$ are internal energy densities, ϵ_ι is specific energy, and velocity $\mathbf{v}(\mathbf{r}, t)$ identical for all massive particles. The equations of state are $P = \sum_\iota P(\rho, \epsilon_\iota)$, $\epsilon_\iota = \epsilon_\iota(\rho, T_\iota)$. The components can exchange energy, can transfer energy by heat conduction not associated with the velocity of the massive particles $m_\iota \neq 0$, and can participate in reactions. Such problems arise in inertial thermonuclear fusion [57], laser ablation experiments [34,131], and astrophysics [84]. This is an intermediate case between the description is based on the Boltzmann equations for the *one-particle distribution function* $f_\iota(\mathbf{r}, \mathbf{p}, t)$ and classical single-component gas dynamics. The basic mathematical problem concerns the gas-dynamic part and the construction of an efficient Godunov-type difference scheme based on an approximate Riemann solver.

Efficient Riemann solvers for such problems are constructed, see, e.g., Ref. [256]. Solvers are also developed for special cases of the two-term equation of state $P = (\Gamma - 1)\rho\epsilon + c_0^2(\rho - \rho_0)$ for a mixture of gases with different temperatures [327] and for a gas mixture assumed to locally satisfy the Mie–Gruneisen equation of state with a linear dependence of pressure on internal energy $\Delta P = \frac{\Gamma_0}{V}\Delta\epsilon$ with single temperature [225].

In the following, an original method based on the *Riemann problem solver* for the multitemperature non-equilibrium gas is described. The method was applied within the plasma physics for the inertial heavy ion fusion [23,57] and for the laser ablation [11] and can be useful for the gravitational collapse taking into account neutrinos transport in astrophysics. Each component in the mixture has its own density and specific internal energy. In the *local model for the equation of state* proposed, it is assumed that the entropy variations in neighboring mesh cells are small and the variations in the dimensionless coefficients of the equations of state (adiabatic indices) $\Gamma_\iota \equiv 1 + P_\iota/(\rho\epsilon_\iota)$ and in the specific internal energies based on the pressure jump across the discontinuity are computed in advance. In the case of an arbitrarily large pressure jump, the model yields physically reasonable results. However, the algorithm for solving the *Riemann problem* was not specially tested for correctness. It was assumed that, due to heat conduction, the Riemann solver deals with small pressure jumps on a fixed grid, but in real cases, the pressure jumps are not small. The SW structure in plasmas was considered in Ref. [278]. This solution is a suitable test for the technique developed.

7.2.1 *Formulation of the problem and the numerical method*

To solve problems in several dimensions, it is convenient to use fixed Euler orthogonal curvilinear coordinates and introduce the concentration $c_\iota = \rho_\iota/\rho$ of substance of sort ι. The system consists of the mass transfer equations for the components,

$$\frac{\partial \rho_\iota}{\partial t} + \mathrm{div}\rho_\iota \mathbf{v} = \rho\dot{c}_\iota, \tag{7.4}$$

the momentum conservation law

$$\frac{\partial \rho \mathbf{v}}{\partial t} + \mathrm{Div}\Pi = 0, \tag{7.5}$$

and the *energy density* equations

$$\begin{aligned}\frac{\partial \rho E_\iota}{\partial t} &+ \mathrm{div}(\rho E_\iota + P_\iota)\mathbf{v} + \mathbf{v}(c_\iota \mathrm{grad}P - \mathrm{grad}P_\iota)\\ &= \mathrm{div}(\kappa \mathrm{grad}T_\iota) + \rho Q_\iota,\end{aligned} \tag{7.6}$$

where the energy densities $E_\iota = \epsilon_\iota + c_\iota \mathbf{v}^2/2$, the tensor $\Pi_{ij} = \rho v_i v_j + P\delta_{ij}$, and the equation of state $P = \sum_\iota P_\iota(\rho, \mathbf{c}, \epsilon_\iota)$ with specific energies $\epsilon_\iota(\rho, \mathbf{c}, \epsilon_\iota)$. The kinetic coefficients $\dot{c}_\iota$, κ_ι, and Q_ι depend on $\rho, \mathbf{c}$, and $\mathbf{T}$ [56]. Operators div$\mathbf{v}$, DivΠ, and grad$P$ are defined in Appendix A. Instead of using the distribution functions $f_\iota(\mathbf{r}, \mathbf{p}, t)$ for the various particles, the description is restricted to gas-dynamic quantities, namely, the component densities $\rho_\iota(\mathbf{r}, t)$ and temperatures $T_\iota(\mathbf{r}, t)$, and assume that the velocities of particles with non-zero masses are identical and equal to $\mathbf{v}(\mathbf{r}, t)$. With the use of this approach, even the transport of photons and fast plasma particles in both transparent and opaque cases can be described with the help of diffusion with flux limiters.

For a single component gas (the sum of equations over ι), *conservation laws* mean that the conservation laws hold in the case of a discontinuous solution. For a *multicomponent gas* in Lagrangian variables (for simplicity in 1D plane case and only the gas-dynamic part: ($m = \int_{r_0}^{r} \rho(\xi, t) d\xi, t$), $\frac{\partial}{\partial t_{\mathrm{E}}} = \frac{\partial}{\partial t_{\mathrm{L}}} - v\rho\frac{\partial}{\partial m}$, $\frac{\partial}{\partial x} = \rho\frac{\partial}{\partial m}$), one has the system of equations

$$\frac{\partial \tau}{\partial t} - \frac{\partial v}{\partial m} = 0, \tag{7.7}$$

$$\frac{\partial c_\iota}{\partial t} = 0, \tag{7.8}$$

$$\frac{\partial v}{\partial t} + \frac{\partial P}{\partial m} = 0, \tag{7.9}$$

$$\frac{\partial E_\iota}{\partial t} + \frac{\partial}{\partial m}(P_\iota v) + v\left(c_\iota \frac{\partial P}{\partial m} - \frac{\partial P_\iota}{\partial m}\right) = 0, \tag{7.10}$$

or the specific energy equation $\frac{\partial \epsilon_\iota}{\partial t} + P_\iota \frac{\partial v}{\partial m} = 0$. The concentrations c_ι are independent of time and are generally not necessary if the initially different substances are distributed over their cells and there are no reactions, i.e., a heterogeneous mixture remains such.

The problem is computed by applying *dimensional splitting*. The heat conduction equations are solved using *central difference* approximations. As a result, the system of partial differential equations is reduced to an ODE system for $\dot{\epsilon}_{\iota,i}$. The differences relate the neighboring cells $\epsilon_{\iota,i-1}$, $\epsilon_{\iota,i}$, and $\epsilon_{\iota,i+1}$. The flux in the heat equation is calculated by approximating the gradient of specific internal energy:

$$F_{\iota,j+1/2} = \left(\kappa_\iota \frac{\partial T_\iota}{\partial x}\right)_{j+1/2}$$
$$= \frac{T_{\iota,j+1} - T_{\iota,j}}{2}\left(\frac{\kappa_{\iota,j}}{x_{j+1/2} - x_{j-1/2}} + \frac{\kappa_{\iota,j+1}}{x_{j+3/2} - x_{j+1/2}}\right). \tag{7.11}$$

The ODE system is solved by applying the implicit *Gear's method* [137], while a matrix of the form $\text{diag}\{1,\dots,1\} + \Delta t \frac{\partial f(Y)}{\partial Y}$ is inverted using *cyclic reduction method* with tridiagonal matrix algorithm. To describe the kinetics of reactions, the ODE system for $\dot{\rho}_\iota$ and $\dot{\epsilon}_\iota$ is solved in each grid cell by employing Gear's method.

The main difficulty is the advection part; the corresponding system of equations is hyperbolic. Assuming that the mixture is homogeneous, the system of equations for a multicomponent gas in Eulerian coordinates strictly corresponds to the differential equations for the same gas in Lagrangian coordinates. The gas-dynamic equations in Eulerian coordinates can also be derived directly from the Boltzmann equations for the distribution functions $f_\iota(\mathbf{r}, \mathbf{p}, t)$, assuming that the massive particles (atoms, ions, and nucleons) have identical velocities and temperatures, while the massless particles (electrons) have their own temperatures. Mathematically, the Lagrangian and Euler equations for the mixture are different. In Lagrangian variables, the multicomponent single-temperature heterogeneous mixture satisfies *conservation laws* as before. The system in Euler variables involves the term $\mathbf{v}(c_\iota \text{grad}P - \text{grad}P_\iota)$, which is different from the divergence of a flux. This term requires a special treatment at discontinuities. The same non-conservative term is involved in the Lagrangian energy equation when the temperatures of the components are different. The term $\text{grad}P_\iota$ at a discontinuity is not defined. *Artificial viscosity* has been introduced by Richtmyer and Morton to describe discontinuities [267].

Actually, the high-order Godunov-type PPM scheme [108] involves artificial viscosity and even a diffusion term for all conservative quantities $\propto \Delta x^2 \max(-\text{div}\mathbf{v}, 0)$, which is quadratic in the mesh size. On *SW*, the order of the scheme is reduced to the first. If the computational errors are summed with different signs, then numerical diffusion is reduced with decreasing mesh size. Thus, the uniqueness of the resulting numerical solution can be verified in practice by mesh refinement. In the classical problem of SW in a plasma, one has only three *conservation laws* and, due to heat conduction, a piecewise smooth temperature, for which a differential equation can be used instead of a conservation law.

The hydrodynamic part of the code is based on a high-order accurate Godunov scheme for single-temperature single-component gas dynamics, see Ref. [108]. In the multidimensional case, fixed Euler coordinates are generally universal and special requirements are imposed on the spatial resolution of the scheme. To generalize the approach, see Ref. [107], one needs an efficient solution algorithm for the *Riemann problem* for the real gas *equation of state* with different temperatures of components.

The data in the *Riemann problem* are prepared using the standard procedure [108]. The system of *conservation laws* is rewritten in *quasilinear form* for the variables $\mathbf{V} = \left\{\tau_\iota \equiv \frac{1}{\rho_\alpha}, u, v, w, P_\alpha\right\}^T$ (here three velocity components are denoted as $\mathbf{v} = (u, v, w)$):

$$\frac{\partial \mathbf{V}}{\partial t} + A\frac{\partial \mathbf{V}}{\partial x} = 0, \tag{7.12}$$

where

$$A_0 = \begin{pmatrix} \ddots & \vdots & \vdots & \vdots & \vdots & \vdots & \vdots \\ \cdots & u & -\tau_\iota & 0 & 0 & 0 & \cdots \\ \cdots & 0 & u & 0 & \tau & 0 & \cdots \\ \cdots & 0 & 0 & u & 0 & 0 & \cdots \\ \cdots & 0 & 0 & 0 & u & 0 & \cdots \\ \cdots & 0 & \tau C_\iota^2 & 0 & 0 & u & \cdots \\ \vdots & \vdots & \vdots & \vdots & \vdots & \vdots & \ddots \end{pmatrix}, \tag{7.13}$$

with squared Lagrangian speed of sound $C_\iota^2 = P_\iota (P_\iota)_{\epsilon_\iota} - (P_\iota)_\tau$. The eigenvalues of the matrix A are real and are given by

$$\lambda_0, \ldots, \lambda_{2\iota_{\max}} = u, \ \lambda_\mp = u \mp c_s, \tag{7.14}$$

where $c_s = \tau C$ is the Euler speed of sound and $C^2 = \sum_\iota C_\iota^2$. Let $\mathbf{l}_\#$ and $\mathbf{r}_\#$ (where the index $\#$ takes the values $0, \ldots, 2\iota_{\max}, \mp$) be the normalized left and right eigenvectors corresponding to the eigenvalue $\lambda_\#$, i.e., $\mathbf{l}_\# A = \lambda_\# \mathbf{l}_\#$, $A\mathbf{r}_\# = \lambda_\# \mathbf{r}_\#$, and $\mathbf{l}_\# \mathbf{r}_{\#'} = \delta_{\#\#'}$. Then *characteristic* equations are obtained by multiplying $\mathbf{l}_\#$ by (7.12): $\mathbf{l}_\# \mathbf{V}_t + \lambda_\# \mathbf{l}_\# \mathbf{V}_x = 0$ or

$$\mathbf{l}_\# \frac{d\mathbf{V}}{dt} = 0 \text{ along the curve } \frac{dx}{dt} = \lambda_\#. \tag{7.15}$$

The variables ρ_ι, $\mathbf{v}$, and P_ι are interpolated in grid cells with the help of a second-degree parabola. This degree is needed to obtain at least the second order on a grid with an arbitrary non-uniform partition. The interpolation procedure eliminates the formation of local minima. Then, time-averaged constant values to the left and to the right of the cell boundaries are determined by solving the system in *quasilinear form* with a constant coefficient matrix A. The computations are similar to those in Ref. [108]. In the case of a smooth flow, there are no discontinuities on the boundaries and values averaged over the time step Δt are obtained on the fixed cell boundaries $x_{j+1/2}$.

The interpolation procedure is as follows [108]:

(1) Determine the time-averaged (over Δt) fluxes on the boundaries of the zones.

- For $a = (\rho_\iota, u, v, w, P, C, \Gamma, \Gamma_\iota^\epsilon \equiv \epsilon_\iota/\epsilon)$, find the zeroth approximations for the initial data in the *Riemann problem*:

$$\tilde{a}_{j+1/2,L} = f^a_{j+1/2,L}(x_{j+1/2} - \tilde{x}_{j+1/2,L}), \tag{7.16}$$

$$\tilde{a}_{j+1/2,R} = f^a_{j+1/2,R}(\tilde{x}_{j+1/2,R} - x_{j+1/2}), \tag{7.17}$$

where

$$f^a_{j+1/2,L}(y) = \frac{1}{y} \int_{x_{j+1/2-y}}^{x_{j+1/2}} a_0(x)dx, \tag{7.18}$$

$$f^a_{j+1/2,R}(y) = \frac{1}{y} \int_{x_{j+1/2}}^{x_{j+1/2}+y} a_0(x)dx, \tag{7.19}$$

$$\tilde{x}_{j+1/2,L} = x_{j+1/2,L} - \max\left(0, \Delta t \left(u_j^n + c_{s,j}^n\right)\right), \tag{7.20}$$

$$\tilde{x}_{j+1/2,R} = x_{j+1/2,L} + \max\left(0, \Delta t \left(u_{j+1}^n + c_{s,j+1}^n\right)\right), \tag{7.21}$$

$c_s = \tau C$ is the Eulerian speed of sound, j is the index of an interval (cell), $x_{j+1/2}$ is the right boundary, and $\Delta x_j \equiv x_{j+1/2} - x_{j-1/2}$. The properties of the interpolating polynomial are as follows: $a_j^n = \frac{1}{\Delta x_j}\int_{x_{j-1/2}}^{x_{j+1/2}} a_0(x)dx$; the function $a_0(x)$ is continuous inside each interval, and $a_{j+1/2} = \frac{d}{dx}\left[\int^x a_0(\xi)d\xi\right]\Big|_{x=x_{j+1/2}}$ must satisfy the identity $a_j \le a_{j+1/2} \le a_{j+1}$. When a_j has a local extremum, the interpolation function must be a constant. Due to these requirements, one can construct a monotonic difference scheme. The formulas for $a_0(o)$, $f_{j+1/2,L}^a$, $f_{j+1/2,R}^a$ can be found in Ref. [108]. The interval $(\tilde{x}_{j+1/2,L}, \tilde{x}_{j+1/2,R})$ at the time t^n includes the region of influence of the solution at the point $(x_{j+1/2}, t^{n+1})$.

- The left and right states for $\mathbf{V}_0 = \{\tau, u, v, w, P\}^T$ in the *Riemann problem* are specified from integration for the corresponding *characteristics* [108] to provide the following relation [107]:

$$\mathbf{V}_{j+1/2,L} \approx \tilde{\mathbf{V}}_{j+1/2,L} + P_{>}\left(\mathbf{V}_j - \tilde{\mathbf{V}}_{j\,j+1/2,L} + \frac{\Delta x_j - \Delta t A_j}{2}\frac{\Delta \mathbf{V}_j}{\Delta x_j}\right), \tag{7.22}$$

$$\mathbf{V}_{j+1/2,R} \approx \tilde{\mathbf{V}}_{j+1/2,R} + P_{<}\left(\mathbf{V}_{j+1} - \tilde{\mathbf{V}}_{j\,j+1/2,R} - \frac{\Delta x_{j+1} - \Delta t A_{j+1}}{2}\frac{\Delta \mathbf{V}_{j+1}}{\Delta x_{j+1}}\right), \tag{7.23}$$

where the operators are $P_{>}(\mathbf{w}) = \sum_{\#,\lambda_{\#}>0}(\mathbf{l}_{\#}\mathbf{w})\mathbf{r}_{\#}$ (having a region of influence on the left) and $P_{<}(\mathbf{w}) = \sum_{\#,\lambda_{\#}<0}(\mathbf{l}_{\#}\mathbf{w})\mathbf{r}_{\#}$ (having a region of influence on the right); the subscript 0 is hereafter omitted. These relations approximate the characteristic equations with allowance for perturbations propagating along the *characteristics*; here $\mathbf{V}_{j+1/2,L}$, $\mathbf{V}_{j+1/2,R}$ are approximate solutions of the characteristic equations to the left and to the right of $x_{j+1/2}$ at the time $t^n + \Delta t/2$.

There is no need to consider the system of quasilinear equations for all independent variables $\left\{\tau_\iota \equiv \frac{1}{\rho_\iota}, u, v, w, P_\iota\right\}^T$

(which is inconvenient) since one has $\rho_\iota/\rho = \text{const}$ on the *characteristic* $dx/dt = u$ and, after $\rho_{j+1/2,\frac{L}{R}}$ was found, one can determine $\rho_{\iota,j+1/2,\frac{L}{R}}$ given the mean values of ρ_ι in the region of influence corresponding to the characteristic $dx/dt = u$.

It turns out that, for determining the increments Γ and Γ_ι^ϵ, it is sufficient to know any initial value (interval-averaged) and the increment of the total *pressure* $P_{j+1/2,\frac{L}{R}} - P_{j,j+1}$. Recall that the *local model for the equation of state* is used for the solution of the *Riemann problem*, see Eqs. (7.30) and (7.29).

(2) For the sets $(\rho_\iota, u, v, w, P, \Gamma, F \equiv c^2\rho/P, \Gamma_\iota^\epsilon)_{j+1/2,\frac{L}{R}}$, solve the *Riemann problem.*
(3) Using the solution of the *Riemann problem* and the sets $(\bar{\rho}_\iota, \bar{u}, \bar{v}, \bar{w}, \bar{P}_i, \bar{\Gamma}_\iota)_{j+1/2}$, construct time-averaged fluxes on the boundaries of the zones and find the solution of the original equations at the time t^{n+1}.

A *local model for the equation of state* simplifies the solution of the *Riemann problem* and makes it possible to obtain fluxes and partial *pressures* of the components in any flow region with discontinuities. Following [107], such model for a multicomponent gas is constructed so as it holds strictly in the case of weak discontinuities. The increment of the specific entropy s across SW is a quantity of the third order of smallness with respect to the *pressure* jump: $\mathcal{O}([P]^3)$. Neglecting the entropy variation behind the SW, one computes the dimensionless coefficients

$$\Gamma_\iota \equiv \frac{P_\iota \tau}{\epsilon_\iota} + 1, \tag{7.24}$$

as functions of the state ahead of the SW and the total *pressure* behind the SW. The local model for the equation of state is used to solve the *Riemann problem* with constant values specified initially to the left and to the right of $x/t = 0$. Moreover, to the left and right of the *contact discontinuity*, the concentrations remain constant, i.e., the *equation of state* is independent of c_ι. The dependence of Γ_ι is very convenient in the construction of a Riemann solver, see Ref. [107]. Using the relation

$$d\epsilon_\iota(s_\iota, \tau) = T_\iota ds_\iota - P_\iota d\tau \tag{7.25}$$

and the assumption

$$ds_\iota = 0, \tag{7.26}$$

one can write

$$\frac{d\Gamma_\iota}{dP_\iota} = \frac{\tau}{\epsilon_\iota} + \frac{P_\iota}{\epsilon_\iota}\frac{d\tau}{dP_\iota} - \frac{P_\iota\tau}{\epsilon_\iota^2}\frac{d\epsilon_\iota}{d\tau}\frac{d\tau}{dP_\iota} = \frac{\tau}{\epsilon_\iota}\left(1 - \frac{\Gamma_\iota}{F_\iota}\right)$$
$$= (\Gamma_\iota - 1)\left(1 - \frac{\Gamma_\iota}{F_\iota}\right)\frac{1}{P_\iota}, \tag{7.27}$$

i.e., one obtains an explicit dependence of the increment of Γ_ι on the increment of the *pressure* of a component. The partial pressure increment depends explicitly on the total pressure increment:

$$dP_\iota = \frac{C_\iota^2}{C^2}dP, \tag{7.28}$$

where the squared Lagrangian speed of sound of a component is given by $C_\iota^2 \equiv -dP_\iota/d\tau = (\partial P_\iota/\partial\epsilon_\iota)P_\iota - \partial P_\iota/\partial\tau$, $P(\epsilon,\tau) = \sum_\iota P_\iota(\epsilon_\iota,\tau)$. In the computations, it is convenient to use the following relations for the fraction of the specific energy of a component:

$$d\Gamma_\iota^\epsilon = \frac{\epsilon d\epsilon_\iota - \epsilon_\iota d\epsilon}{\epsilon^2} = \frac{(\epsilon P_\iota - \epsilon_\iota P)dP}{\epsilon^2 C^2}$$
$$= \frac{(\epsilon P_\iota - \epsilon_\iota P)\tau dP}{F P\epsilon^2} = \Gamma_i^\epsilon\frac{\Gamma_\iota - \Gamma}{F}\frac{dP}{P}. \tag{7.29}$$

With the use of the notation $F_\iota \equiv C_\iota^2\tau/P_\iota$, $F \equiv C^2\tau/P$, and $d\epsilon_\iota = -P_\iota d\tau = -P_\iota\frac{d\tau}{dP_\iota}dP_\iota = \frac{P_\iota dP_\iota}{C_\iota^2} = \frac{P_\iota dP}{C^2}$, the increment of the dimensionless variable $\Gamma \equiv \sum_\iota \Gamma_\iota\epsilon_\iota/\sum_\iota\epsilon_\iota$ is given by (see details of the transformations in Ref. [11])

$$d\Gamma = \sum_\iota \frac{\epsilon_\iota}{\epsilon}(\Gamma_\iota - 1)\left(1 - \frac{\Gamma_\iota}{F_\iota}\right)\frac{F_\iota}{F}\frac{dP}{P} + \sum_\iota\frac{\Gamma_\iota P_\iota dP}{\epsilon C^2} - \sum_\iota\frac{\Gamma_\iota\epsilon_\iota P dP}{\epsilon^2 C^2}$$
$$= (\Gamma - 1)\left(1 - \frac{\Gamma}{F}\right)\frac{dP}{P}, \tag{7.30}$$

as expected for a single-component gas, see Ref. [107], or for a single component (7.27). A key point in this work and the main difference of the proposed gas-dynamic approach for multicomponent multitemperature gases from [107,108] lie in formula (7.29) and in the derivation of formula (7.30) for a multicomponent gas.

The assumption that the entropy variation is negligibly small is used to compute the variations in the dimensionless coefficients. However, this assumption is not used in the computation of other parameters, specifically those behind the SW.

The solution of the *Riemann problem* consists of a *contact discontinuity*, on which the *pressure* and the normal velocity are continuous,

$$u_{*L}(P_*) - u_{*R}(P_*) = 0, \tag{7.31}$$

and of waves of two types propagating to the left and to the right, namely, SW or a *rarefaction wave* (here, the star denotes the state behind the SW or rarefaction wave).

The *Rankine–Hugoniot conditions* at the discontinuities are given by

$$[\rho_i v_n] = 0, \tag{7.32}$$

$$[\rho v_n^2 + P] = 0, \tag{7.33}$$

$$[\rho v_n v_t] = 0, \tag{7.34}$$

$$\left[v_n\left(\frac{\rho(v_n^2+v_t^2)}{2} + \rho\epsilon + P\right)\right] = \rho v_n\left[E + \frac{P}{\rho}\right] = 0. \tag{7.35}$$

(1) *Tangential discontinuity*:

$$\rho v_n, [P] = 0, \tag{7.36}$$

v_t and c_ι have arbitrary jumps.

(2) SW:

$$\rho v_n, [\tau] \neq 0, \tag{7.37}$$

$c_\iota = 0$, $[v_t] = 0$.

In what follows, $u \equiv v_n$. At the SW,

$$\pm W[\tau] + [u] = 0, \tag{7.38}$$

$$\pm W[u] - [P] = 0, \tag{7.39}$$

$$\pm W[E] - [uP] = 0, \tag{7.40}$$

$$[c_\iota] = 0, \tag{7.41}$$

where the mean Lagrangian velocity of the wave is introduced

$$W_S = \frac{|P_* - P_s|}{|u_{*S} - u_S|}, \tag{7.42}$$

and the sign depends on the direction $(\pm - \binom{R}{L})$.

To determine P_* at the *contact discontinuity*, one can use the efficient *Newton's iterations*, see Ref. [108],

$$u^{\nu}_{*S} = u_S \pm \frac{P^{\nu}_* - P^{\nu}_S}{W^{\nu}_S}, \tag{7.43}$$

$$P^{\nu+1}_{*S} = P^{\nu}_{*S} - \frac{Z_L Z_R(u^{\nu}_{*R} - u^{\nu}_{*L})}{Z_L + Z_R}, \tag{7.44}$$

where $Z_S = |dP_*/du_{*S}|$ and $S = L, R$.

For the *rarefaction wave*,

$$\frac{dP_*}{du_{*S}} = C(P_*, \rho_*), \tag{7.45}$$

and, for the SW,

$$\frac{dP_*}{du_{*S}} = \frac{W^2}{W - (dW/dP_*)[P]}, \tag{7.46}$$

where

$$\frac{dW^2}{dP_*} = \frac{(C^2 - W^2)W^2}{((P_S + P_*)/(2\partial P/\partial \epsilon) - \partial P/\partial \tau)[P]}, \tag{7.47}$$

as obtained in Ref. [107] by differentiating $[P^2]/2 = W[\epsilon]$ and $\epsilon_* = \epsilon(P_*, \tau - [P]/W)$ with respect to P_*. To simplify the computations, it was suggested in Ref. [108] that the same formulas as for SW are used to determine the state from the given *pressure* P_* behind the *rarefaction wave*. The values inside the rarefaction wave between $u \pm c_S$ and $u \pm c_{S*}$ can be computed using *linear interpolation* with respect to the self-similar variable x/t. As a result, the integration in the *rarefaction wave* from P_S to P_* is avoided. In the case of a *local model for the equation of state*, it was noted in Ref. [107] that the SW computations can be simplified using $\Gamma_{S*}(P_*)$ known *a priori* from the local *equation of state*. An explicit formula for W can be derived from $W^2[\epsilon] = [P^2]/2$ and $\epsilon_{S*} = \epsilon\left(P_*, \tau_S - \frac{[P]}{W_S^2}\right)$:

$$W^2 = \frac{P_*^2 - P^2}{2} \Bigg/ \left[\frac{P_*}{\Gamma_* - 1}\left(\tau - \frac{P_* - P}{W^2}\right) - \epsilon\right]. \qquad (7.48)$$

Given P_*, one determines Γ_{*S} (7.30) and Γ^ϵ_{*S} (7.29), and the approximate value W_S^2 from (7.48) is used to compute $\tau_{S*} = \tau_S - \frac{[P]}{W_S^2}$. Then one computes W_S from (7.42), $u_{S*} = u_S \pm (P_* - P_S)/W_S$, $\epsilon_{S*} = \epsilon_S - (\tau_{S*} - \tau_S)(P_* + P_S)/2)$ $\rho_{\iota,S*} = \rho_{\iota,S}\tau_{S*}/\tau_S$, and $\epsilon_{\iota,*S} = \Gamma^\epsilon_{*S}\epsilon_{S*}$. One has ρ_{S*}, ϵ^ι_{S*} and knows the necessary derivatives for

$$Z_S = \frac{W_S}{1 - \frac{(C_*^2 - W_S^2)/2}{C_S^2 + \partial P/\partial\epsilon*(P_S - P_*)/2}} \qquad (7.49)$$

and $P_{\iota*}(\tau_{S*}, \epsilon_{S*})$ from the *equation of state*. The next iteration can be performed to determine the *pressure* P_* on the *contact discontinuity*. Due to the approximation used, one can avoid the evaluation of the integrals at every iteration step for P_* in the *rarefaction wave* and the solution of additional nonlinear equations with approximate value (7.48) in the computation of τ_{S*} with an *a priori* known fraction of the internal energy $\Gamma^\epsilon(P_*)$ (7.30).

On a fixed Eulerian grid, the values of $\rho_\iota, u_\iota, P_\iota$, and Γ^ϵ_ι on the boundary of the intervals (i.e., at the point $x/t = 0$) and the position of the interface relative to $x/t = 0$ are needed to determine the tangential velocity on the interface in multidimensional problems.

The *local model for the equation of state* proposed resolves the uncertainty occurring when the specific internal energy and the *pressure* of a mixture component behind the SW (*rarefaction wave*) are computed from known values of $\Gamma^\epsilon_\iota \equiv \epsilon_\iota/\epsilon$ behind the wave. When the energy equations for all components are added, the resulting difference scheme is conservative if the normalization $\sum_\iota P_{\iota*} = P_*$ is used. The "adiabatic model" for the *equation of state* is related to the dimensionless coefficients Γ (7.30) and Γ^ϵ_ι (7.29). The entropy jumps occurring behind the SW are non-zero.

7.2.2 *Shock waves in hydrogenous plasma*

As a test, consider an SW in hydrogen arising if a constant velocity is given at the right-hand boundary and the gas moves in an immovable hydrogenous plasma. The initial state is completely ionized hydrogen. The system contains protons and electrons. Electrons do not contribute to the mass but they transport the heat by heat conduction and exchange their energy with protons. The temperatures differ near SW. The kinetic coefficients are taken from [278]: The velocity of the energy exchange between components in Eq. (7.6) is

$$\rho \chi_{ie}(T_i - T_e) = \frac{3m_e}{m_p}\frac{n_e}{\tau_e}(kT_e - kT_i), \tag{7.50}$$

where $\tau_e = 3\sqrt{m_e}(kT_e)^{3/2}/(4\sqrt{2\pi}\lambda e^4 n_e)$, the Coulomb logarithm is $\lambda = \frac{1}{2}\ln\left(\frac{kT_eT_i}{T_e+T_i}\right)^3 \frac{1}{e^6 n_e}$, $n_e = n_i = \rho/m_p$, and the heat flow in the heat conduction of the electrons is $-\kappa_e \nabla(kT_e)$ with the thermal conduction coefficient

$$\kappa_e = 3.16 n_e k T_e \tau_e / m_e. \tag{7.51}$$

The equation of state is the ideal monatomic gas for protons and electrons with $\gamma_{i,e} = 5/3$. Its electroneutrality is assumed and it is assumed that the concentrations and velocities of the protons and electrons are the same everywhere. We neglect the jump of the density of electrons on SW. In Ref. [278], the structure of a stationary SW is computed from the conditions of the jump and solutions of ordinary differential equations in the regions before the SW and behind its front.

The initial pressure of the hydrogen is approximately equal to the atmospheric pressure and the initial temperature is selected from the assumption of complete ionization. It turns out that if the initial temperature is 10^4 K and the initial density is 10^{-6} g $\cdot$ cm^{-3}, then the hydrogen is half-ionized. The Saha equations obtained if the sums of the chemical potentials for the initial and final particles are equated to each other have the form

$$\frac{n_e n_p}{n_H} = \frac{(m_e kT)^{3/2}}{(2\pi\hbar)^3} g_e e^{-I/(kT)}, \tag{7.52}$$

where the ionization potential $I = 13.6$ eV. It turns out that the hydrogen is ionized at a low temperature satisfying the condition $kT \ll I$. The velocity of the injected gas at the right-hand boundary is selected to obtain a strong stationary SW such that the density jump on it is close to four: In that case, it is possible to compare the numerical solution with the Shafranov solution for a strong SW. The gas velocity at the right-hand boundary is assigned to be equal to $-4 \cdot 10^7$ cm $\cdot$ s^{-1}, while the pressure and temperature of the components are assigned to be the same as in the unperturbed gas. This yields the gas velocity $-2 \cdot 10^7$ cm $\cdot$ s^{-1}, the gas pressure $5.37 \cdot 10^8$ dyn $\cdot$ cm^{-2}, and the temperature $8.11 \cdot 10^5$ K behind the SW. The pressure of the blackbody radiation with that temperature is $\frac{4}{3}\frac{\pi^2}{60(\hbar c)^3}(kT)^4 = 1.2 \cdot 10^9$ dyn $\cdot$ cm^{-2} which exceeds the gas pressure. Thus, for a correct formulation (from the point of view of physics), we have to either introduce photons and take into account their energy transport energy exchange with electrons or decrease the velocity on the right-hand boundary and the intensity of SW, which is done in this chapter.

The numerical solution of the problem is displayed in Fig. 7.1, and the parameters of the SWs are presented in Table 7.1. The mesh $\Delta x_j = 2.5 \cdot 10^{-3}$ cm is used. As far as the gas moves, a stationary SW moving over the unperturbed gas is formed. The second SW

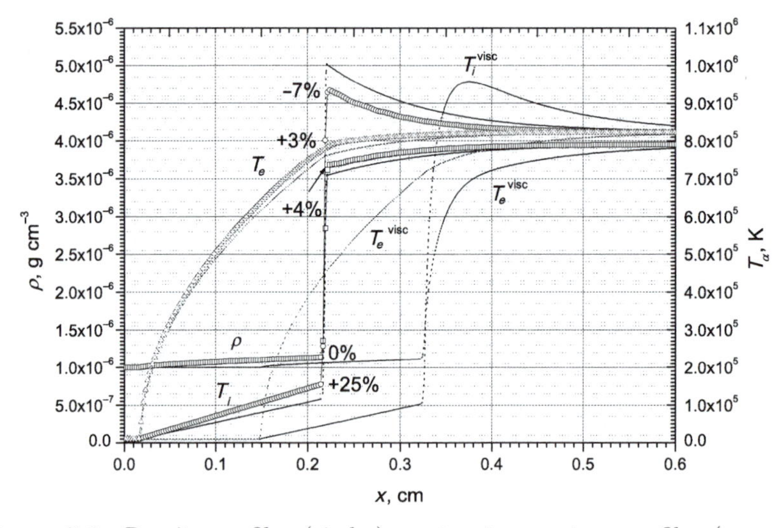

Figure 7.1 Density profiles (circles), proton temperature profiles (squares), electron temperature profiles (triangles), and comparison with exact solution of Shafranov's problem (dotted lines) for strong SW for case where $M = 16$. Figures near SW denote relative accuracy of numerical solution. Solution taking into account physical viscosity of protons is displayed as well (dotted lines marked as "visc", which are displaced to the right along the x-axis).

Table 7.1 Density and temperature parameters in Shafranov's problem on strong SWs with $M = 16$, propagating over unperturbed gas with $\rho_0 = 1 \cdot 10^{-6}$ g cm^{-3}, $T_0 = 1 \cdot 10^4$ K.

	ρ_1/ρ_0	ρ_2/ρ_0	T_{i1}/T_0	T_{i2}/T_0	T_e/T_0
Shafranov solution	1.13	3.53	13.0	101	55.4
Numerical solution	1.13	3.68	15.3	93.4	78.0
Exact solution of problem with zero viscosity	1.13	3.55	11.6	101	76.1
Exact solution of problem with physical viscosity	1.13	3.46	10.7	95.7	75.0
Numerical solution of problem without heat conduction	1.00	3.96	1.00	157	1.00...8.05
Exact solution with zero viscosity	1.00	3.96	1.00	135	1.00...26.7

from the right moving over the gas with velocity $v = -4 \cdot 10^7$ cm$\cdot$s^{-1} is of no interest. Near the SW, the profiles of all values are displaced with the constant velocity of the front of the SW and remain unchanged. The state of the plasma is not in equilibrium near the SW; equilibrium is established at a distance behind the front. A jump of the density of the protons and a jump in their temperature take place on the SW. Due to the heat conduction of electrons, their temperature T_e is continuous and piecewise smooth. To define the remaining values in the discontinuity region well, the conservation laws for the mass, momentum, and total energy are obeyed even without use of the artificial viscosity. The sound speed in the unperturbed gas is $c_0 = \sqrt{(5/3)P_0/\rho_0} = 1.66 \cdot 10^6$ cm $\cdot$ s^{-1}, i.e., the Mach velocity of the SW is $M = |D|/c_0 = 16$ with respect to the unperturbed gas. To compute the velocity D of the SW, we use the preservation of the flow of particles in the reference frame of the SW: $j = nv_L = \text{const} = (u_i - D)u_i$, where v_L is the velocity in the reference frame of the SW, u is the velocity in the immovable laboratory frame, and the index takes the values L and R, which are the state before and after the SW, respectively. In Fig. 7.1, the coordination of the solution of the problem with the exact one is exact up to fractions of a percent. Note that, in spite of the heat conduction, the pressure jump on SW is not small: On three intervals, it is almost equal to one order of magnitude, while the pressure difference between neighboring computation intervals is tripled. Thus, the proposed method to solve the Riemann problem yields a physically reasonable result in spite of the assumption that the entropy jumps under the computation of dimensionless coefficients of gases.

Shafranov's solution for the structure of a strong SW in two-component hydrogenous plasma seemed to be a successful test of the technique. For a strong SW, the jumps of the temperatures and density are known. In Ref. [11], that solution was used to test the developed technique. The following two conclusions are derived: (1) The jumps obtained by the integrating technique for the system of two-temperature gas dynamics and the Shafranov solution are different from each other. (2) Shafranov's solution from [278] is not exact. Shafranov worked with relations on discontinuities and solved a system of ODEs from both sides of the discontinuity. According to Shafranov (regarding the piecewise-smooth temperature of electrons on an SW), $\left(-\frac{\partial T_e}{\partial x}\right)_1 > \left(-\frac{\partial T_e}{\partial x}\right)_2$, and, therefore, heat flows are different; further, $\left(-\kappa\frac{\partial T_e}{\partial x}\right)_1 > \left(-\kappa\frac{\partial T_e}{\partial x}\right)_2$, and the dependence of the thermal conduction coefficient on the concentration disappears. Heat flows caused by the heat conduction of electrons are to be included in the relations on discontinuities; moreover, we have to know the profile of the temperature of electrons from both sides of the discontinuity. In the developed scheme, the jump of the temperature of ions in SWs is lower than Shafranov's solution, but Shafranov's solution also underestimates this jump because it does not take into account the heat conduction in SWs. Thus, Ref. [11] leads to the conclusion that the scheme works correctly but the quantitative measure of the scheme's quality is not determined.

To estimate the abilities of the method, it is natural to require obtaining Shafranov's solution. According to Shafranov, different velocities and charge separation are admissible in an SW itself. Within our electroneutrality assumption, the velocities of protons and electrons coincide everywhere.

7.2.3 *The solution of Shafranov's problem*

The easiest way to obtain the solution of the Shafranov problem is to solve the system of partial differential equations. To exclude discontinuities, we can introduce the viscosity protons, i.e., pass from the hyperbolic system of equations to a parabolic one. The mass conservation law is

$$\frac{\partial \rho}{\partial t} + \frac{\partial \rho v}{\partial x} = 0.$$

The momentum conservation law is

$$\rho\left(\frac{\partial v}{\partial t} + v\frac{\partial v}{\partial x}\right) + \frac{\partial P}{\partial x} = \frac{\partial}{\partial x}\left[\frac{4\eta_p}{3}\left(\frac{\partial v}{\partial x}\right)\right]$$

or

$$\frac{\partial \rho v}{\partial t} + \frac{\partial}{\partial x}\left(\rho v^2 + P - \frac{4\eta}{3}\frac{\partial v}{\partial x}\right) = 0.$$

The equations for the energies of the components are as follows:

$$\rho\left(\frac{\partial \epsilon_p}{\partial t} + v\frac{\partial \epsilon_p}{\partial x}\right) = -\left(P_p - \frac{4\eta}{3}\frac{\partial v}{\partial x}\right)\frac{\partial v}{\partial x} - \chi\left(k_{\mathrm{B}}T_p - k_{\mathrm{B}}T_e\right),$$

$$\frac{\partial \rho E}{\partial t} + \frac{\partial}{\partial x}\left(\left(\rho E + P - \frac{4\eta}{3}\frac{\partial v}{\partial x}\right)v - \kappa\frac{\partial k_{\mathrm{B}}T}{\partial x}\right) = \rho Q,$$

$$\rho\left(\frac{\partial \epsilon_e}{\partial t} + v\frac{\partial \epsilon_e}{\partial x}\right) = -P_e\frac{\partial v}{\partial x} + \frac{\partial}{\partial x}\left(\kappa_e\left(\frac{\partial k_{\mathrm{B}}T_e}{\partial x}\right)\right) - \chi\left(k_{\mathrm{B}}T_e - k_{\mathrm{B}}T_p\right).$$

The necessary kinetic coefficients are as follows:

$$k_{\mathrm{B}}T_\alpha = (\gamma - 1)\, m_p \epsilon_\alpha,$$

$$\kappa_e \sim \frac{3.16 n k_{\mathrm{B}}T_e}{m_e}\tau_e = \frac{3.16}{\sqrt{m_e}}\frac{3\left(k_{\mathrm{B}}T_e\right)^{5/2}}{4\sqrt{2\pi}\lambda e^4},$$

$$\tau_e = \frac{3\sqrt{m_e}\left(k_{\mathrm{B}}T_e\right)^{3/2}}{4\sqrt{2\pi}\lambda e^4 n},$$

$$\lambda = \frac{1}{2}\ln\left[\left(\frac{k_{\mathrm{B}}T_p k_{\mathrm{B}}T_e}{k_{\mathrm{B}}T_p + k_{\mathrm{B}}T_e}\right)^3\frac{1}{e^6 n}\right],$$

$$\kappa_p \sim \frac{1}{\sigma}\sqrt{\frac{k_{\mathrm{B}}T}{m_p}},$$

$$\eta_p \sim \frac{1}{\sigma}\sqrt{m_p k_{\mathrm{B}}T} = \frac{\kappa_p}{\sqrt{\frac{k_{\mathrm{B}}T}{m_p}}}\sqrt{m_p k_{\mathrm{B}}T} = \kappa m_p = m_p\frac{3.92}{\sqrt{m_p}}\frac{3k_{\mathrm{B}}\left(k_{\mathrm{B}}T_i\right)^{3/2}}{4\sqrt{2\pi}\lambda e^4},$$

$$\chi_e = \frac{3m_e}{m_p}\frac{n}{\tau_e} = \frac{3m_e}{m_p} n \frac{4\sqrt{2\pi}\lambda e^4 n}{3\sqrt{m_e}\,(k_\mathrm{B}T_e)^{3/2}},$$

$$\sigma_{ij} = \rho v_i v_j + P\delta_{ij} - \sigma'_{ij},$$

$$\sigma_{xx} = \rho v^2 + P - \sigma'_{xx} = \rho v^2 + P - \frac{4\eta}{3}\frac{\partial v}{\partial x},$$

$$\sigma'_{xx} = \frac{4\eta}{3}\frac{\partial v}{\partial x} + \zeta\frac{\partial v}{\partial x} = \frac{4\eta}{3}\frac{\partial v}{\partial x}.$$

To solve the problem numerically, it is easier to use the Lagrange variables $(dm = \rho dx, t)$ moving together with the substance. We introduce the mesh $m_{j+1/2}$ and the values $v_{j+1/2}$ on the boundaries and the values $\epsilon_{p,j}$, $\epsilon_{e,j}$ on the intervals of the mesh. Further, the method of lines is used: we leave the derivatives with respect to time and use central differences to approximate spatial derivatives. The equation for the Lagrange coordinate is

$$\frac{dr_{j+1/2}}{dt} = v_{j+1/2}, \rho_j = \frac{\triangle m_j}{\triangle r_j},$$

$$\frac{dv_{j+1/2}}{dt} + \frac{\partial P}{\partial m} = \frac{\partial}{\partial m}\left[\frac{4\eta_p}{3}\left(\frac{\partial v}{\partial r}\right)\right],$$

the equation for the velocity is

$$\frac{d\epsilon_{p,j}}{dt} = -\left(P_p - \frac{4\eta}{3}\frac{\partial v}{\partial x}\right)\frac{\partial v}{\partial m} - \frac{\chi}{\rho}\left(k_\mathrm{B}T_p - k_\mathrm{B}T_e\right),$$

and the equations for the specific energies are

$$\frac{d\epsilon_{e,j}}{dt} = -P_e\frac{\partial v}{\partial m} + \frac{\partial}{\partial m}\left(\kappa_e\left(\frac{\partial k_\mathrm{B}T_e}{\partial r}\right)\right) - \frac{\chi}{\rho}\left(k_\mathrm{B}T_e - k_\mathrm{B}T_p\right).$$

The system of ODEs is solved by the implicit Gear method (see Ref. [137]). If the physical viscosity is taken into account, then the width of the viscous jump is smaller by a factor of $\sqrt{m_p/m_e}$ than the width of the region of the electron heat conduction.

Figure 7.1 displays the numerical solution of the problem on SWs in plasma, the exact solution of the mathematical problem, and the solution of the physical problem with the genuine viscosity. The differences of the jumps in the developed technique and in the solution of the mathematical problem are marked. The physical problem is closer to the numerical solution. The width of the viscous jump is smaller by a factor of $\sqrt{m_p/m_e}$ than the width of the jump of the heat conduction of electrons and it is clearly solved on the detailed mesh. For a strong SW, there is a strong coincidence of all values in the gas-dynamic code. For values of the density and temperatures, the difference is about several percent; the only exception is the 25th total pre-SW pressure is small.

However, we select a strong SW: The pressure after the SW is three orders of magnitude higher than the pressure before the SW. We can reduce the intensity of the SW to ensure the approximate model of the equation to be formally applicable. For the post-SW pressure jump of one order of magnitude and $M = 3.5$, the computation is more than acceptable for the density values (see Fig. 7.2) and temperature values: The worst case is the temperature jump (deficit) of 5% for the protons behind the SW (see Fig. 7.3).

The problem where the heat conduction is turned off (in principle, it can be implemented by the transverse magnetic field) is also

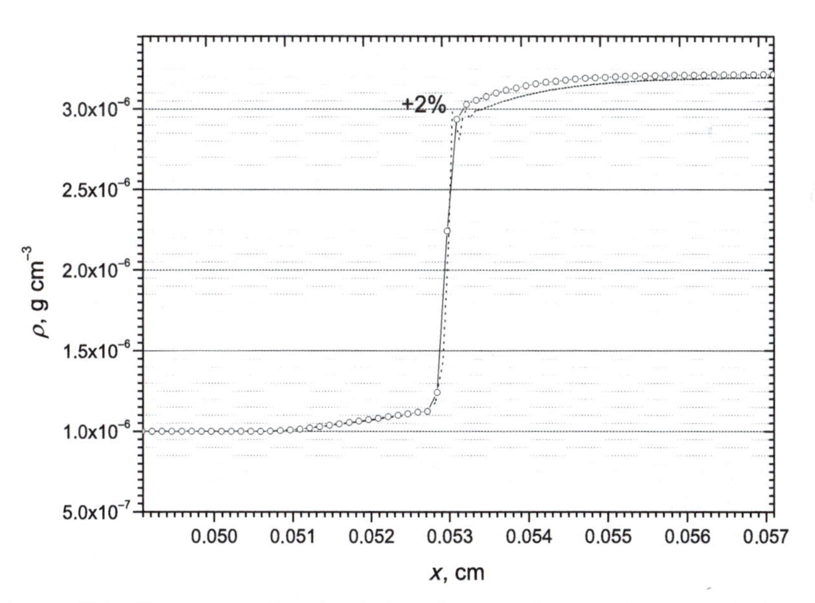

Figure 7.2 Density profiles (circles) and comparison with exact solution of Shafranov's problem (dotted line) for weak SW for case where $M = 3.5$. Figures near SW demonstrate better relative accuracy of numerical solution.

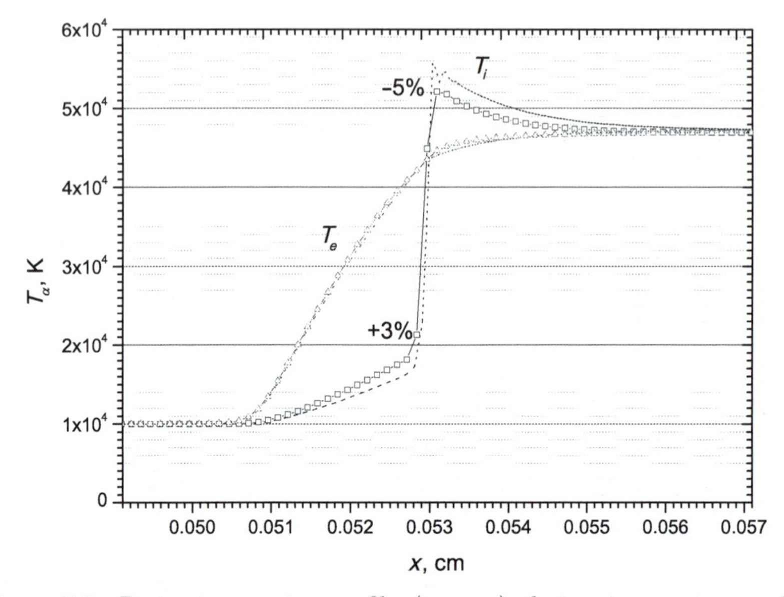

Figure 7.3 Proton temperature profiles (squares), electron temperature profiles (triangles), and comparison with exact solution of Shafranov's problem (primes) for weak SW for case where $M = 3.5$.

interesting for the verification of the method. Is it possible to obtain a solution by the proposed technique? For the case where the SW is weak, the assumption about a small variation of the entropy is applicable and the solution is uniquely defined. The increase in the velocity of the injected gas leads to the growth of the proton temperature behind the SW and to the electron temperature tending to a constant. The total pressure and velocity have constant values everywhere apart from two intervals on the SW. Formally, we have three conservative values and three equations in the form of the law of conservation. For the temperature difference, we have a differential equation. The question is whether the computation of the temperature difference behind the SW is correct. Simple reasoning leads to the conclusion that all the heat energy behind a strong SW falls on protons; further, we have a temperature equalization region. Thus, the T_i behind a strong SW has to be double its value in the temperature equalization region, i.e., approximately $\sim 1.6 \cdot 10^6$ K. If time is small, the relaxation term on the right-hand part is small and, in fact, we obtain a numerical solution of the Riemann problem. We have the correct jump of the ion temperature because the entire internal energy falls on protons behind the strong SW (see Table 7.1). The used approximate solver of the Riemann problem yields equal temperatures for protons and electrons everywhere after the SW provided that they are equal to each other before the wave: $\gamma_i^\epsilon = \gamma_e^\epsilon$. Formally, the solver for a strong pressure jump (almost three orders of magnitude per two mesh intervals in the problem without heat conduction) is not correct. From it, we are required to provide a physically reasonable solution and a conservative scheme for the total energy of all the components. The correct pressure jump for protons is caused by the scheme's viscosity on the SWs (the decrease of the order of magnitude of the method for a large pressure jump until the first one) and the electron pressure gradient in the equations for energies (see Table 7.1). The same table presents the exact solution of the system of equations with the underestimated physical-viscosity coefficient (with factor 0.02). Even this artificially decreased width of the transitional region erodes the jump effect of the ions' temperature: The parameters obtained in the gas-dynamic scheme (the mesh $\Delta x = 1.5 \cdot 10^{-5}$ cm) are closer to the idealized model without viscosity and heat conduction (on the mesh $\Delta x = 2.5 \cdot 10^{-3}$ cm).

7.2.4 *Shock wave in hydrogen plasma*

As a test, an *SW* in hydrogen arising in a gas flowing in a hydrogen plasma at rest when a constant velocity is specified on the right boundary is considered. Initially, the hydrogen is completely ionized. The system involves protons and electrons. The electrons are treated as massless, but they transfer heat by conduction and exchange energy with the protons. The temperatures near the *SW* are different. The kinetic coefficients are specified as in Ref. [278]: The rate of energy exchange in energy equations (7.6) is

$$\rho\chi_{ie}(T_i - T_e) = \frac{3m_e}{m_p}\frac{n_e}{\tau_e}(kT_e - kT_i), \tag{7.53}$$

where $\tau_e = 3\sqrt{m_e}(kT_e)^{3/2}/(4\sqrt{2\pi}\lambda q^4 n_e)$; the *Coulomb logarithm*

$$\lambda = \frac{1}{2}\ln\left(\frac{kT_eT_i}{T_e + T_i}\right)^3 \Big/ (q^6 n_e), \tag{7.54}$$

$n_e = n_i = \rho/m_p$, and the heat flux in electron heat conduction is $-\kappa_e \nabla(kT_e)$ with thermal conductivity

$$\kappa_e = 3.16 n_e k T_e \tau_e / m_e. \tag{7.55}$$

As an *equation of state*, an ideal *monoatomic gas* for protons and electrons with $\Gamma_{i,e} = 5/3$ is used. The *electroneutrality condition* is assumed to hold, the concentrations and velocities of the electrons and protons everywhere coincide, and the jump in the electron density on the SW is neglected. Shafranov [278] calculated the structure of a stationary SW by using the discontinuity conditions and solving the ordinary differential equations ahead of and behind the SW front.

The initial hydrogen *pressure* is approximately equal to the atmospheric one. The initial temperature is chosen so as to achieve complete ionization. It turns out that the hydrogen is half ionized when the initial temperature is 10^4 K and the density is 10^{-6} g $\times$ cm^{-3}. The *Saha ionization equation* (derived by equating the initial and final sums of the *chemical potentials* of the particles) for hydrogen is

$$\frac{n_e n_p}{n_H} = \frac{(m_e kT)^{3/2}}{(2\pi\hbar)^3} g_e e^{-I/(kT)}, \tag{7.56}$$

where the *ionization potential* is $I = 13.6$ eV. The hydrogen is ionized at rather low temperature $kT \ll I$. The velocity of the gas injected at the right boundary is chosen so as to obtain a strong stationary SW on which the density jump is equal to 4 with a sufficient degree of accuracy in order to compare the numerical solution with Shafranov's one for a strong SW. The gas velocity on the right boundary is set to -4×10^7 cm $\times$ s^{-1}, while the *pressures* and temperatures of the components are as in the unperturbed gas. Behind the SW, the gas velocity is -2×10^7 cm $\times$ s^{-1}, the pressure is 5.37×10^8 din $\times$ cm^{-2}, and the temperature is 8.11×10^5 K. The black body radiation pressure with this temperature is $\frac{4}{3}\frac{\pi^2}{60(\hbar c)^3}(kT)^4 = 1.2 \times 10^9$ dyn $\times$ cm^{-2} and exceeds the gas pressure. Therefore, for a correct physical formulation of the problem, one needs to introduce photons and take into account their energy transfer and energy exchange with electrons or, alternatively, the velocity on the right boundary and the

SW intensity have to be reduced, in which case one has no exact solution to be compared with.

The numerical solution of the problem is shown in Fig. 7.1. A grid with $\Delta x_j = 5 \times 10^{-3}$ cm is used. In the course of motion of the gas, a stationary SW is formed, propagating relative to the unperturbed gas. Another SW (on the right) moves relative to the gas at the velocity $v = -4 \times 10^7$ cm $\times$ s^{-1}, but it is of no interest. In the figures, the profiles of all quantities near the SW shift at a constant velocity (of the SW front) and remain unchanged. The plasma is in a non-equilibrium state near the SW, but equilibrium is established at some distance behind the SW front. Jumps in the proton density and temperature are observed on the SW. Due to electron heat conduction, the electron temperature T_e is continuous and piecewise smooth. The other quantities near the discontinuities can be determined from the mass, momentum, and total energy *conservation laws* without using *artificial viscosity*. The speed of sound in the unperturbed gas is $c_0 = \sqrt{(5/3)P_0/\rho_0} = 1.66 \times 10^6$ cm $\times$ s^{-1}, i.e., the *Mach speed* of the SW is $M = |D|/c_0 = 16$ relative to the unperturbed gas.

In Fig. 7.1, the solution of the problem agrees with Shafranov's solution within several percent. Interestingly, despite the heat conduction, the *pressure* jump on the SW is not small. Specifically, it reaches nearly one order of magnitude over three mesh spacings and the pressure values in neighboring mesh spacings differ by three times. Thus, the Riemann solver proposed yields a physically interpretable result despite the assumption for the entropy jump smallness used in computation of the dimensionless coefficients of the gases.

A remark has to be made about the noticeable difference in the solutions for the proton temperature behind the SW. Shafranov admitted the different velocities and the separation of charges in the SW. Within our electroneutrality approximation, the proton and electron velocities everywhere coincide, and the jump in the proton temperature on the SW has to be even higher. The relations between the independent variables on opposite sides of the discontinuity $j = nv_L = \text{const}$, $\frac{v_L^2}{2} + \Gamma\epsilon_i = \frac{v_L^2}{2} + \frac{\Gamma k_B T_i}{(\Gamma-1)m_p} = \text{const}$ ($v_L = u - D$ is the velocity in the reference frame of the SW) were verified in Ref. [11]. The conservations of the mass flux and energy flux in the calculations was satisfactory.

7.2.5 *Homogeneous motion of a gas mixture with contact discontinuities*

The numerical solution for homogeneous motion of a mixture of ideal gases with different particle masses and identical concentrations is shown in Fig. 7.4. One has two boundaries separating three ideal gases with constant *pressure*, velocity $v = 1$, identical constant temperatures, and identical concentrations $n_\iota = \rho_\iota/m_\iota$. The masses of the particles differ by several orders of magnitude: $m_\iota = 1, 10, 100$. The test demonstrates that the high-order accurate Godunov scheme excellently preserves the interfaces. The blur of the concentrations remains within three intervals. The velocities, densities, and temperatures remain unperturbed. This means that certain gas flows in regions of complex geometry can be simulated on structured meshes by applying shock-capturing methods with introducing concentrations and tabulated equations of state. This approach can be very convenient in the case of instabilities developing on the boundaries.

The nonlinear heat conduction of Shafranov's solution demonstrates the remarkable property that the heat conduction wavefront propagates at a finite velocity, in contrast to the infinitely high velocity of heat propagation in linear heat conduction. Diffusion transfer

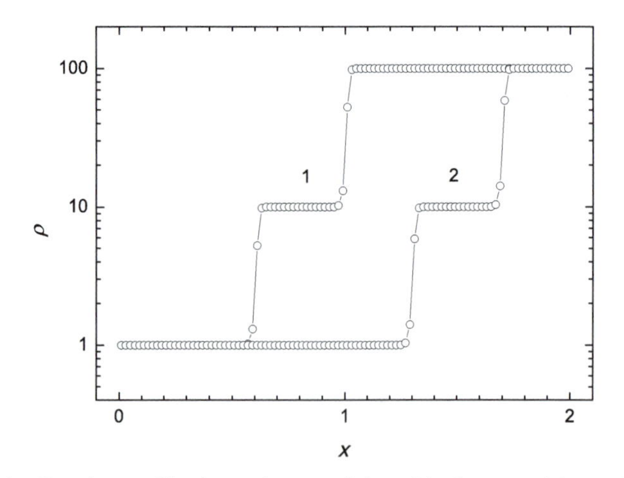

Figure 7.4 Density profiles in a mixture of three ideal gases with particle masses 1, 10, and 100 for the constant velocity $v = 1$, constant *pressure*, and identical constant temperatures and concentrations at the times $t = 0.11$ (1), 0.81 (2). Dimensionless variables were used in the problem.

is qualitatively described using a nonlinear diffusion coefficient with introducing (if necessary) a flux limiter at the interval boundary $j + 1/2$:

$$F_{\iota,j+1/2} \longrightarrow \frac{F_{\iota,j+1/2}}{|F_{\iota,j+1/2}|/F_{\iota\max,j+1/2} + 1}. \tag{7.57}$$

For photons, for example, the maximum flux is given by $F_{\gamma\max} = c\rho\epsilon_\gamma$.

It is necessary to make a remark about the application of the described multitemperature code for the *gravitational collapse* with the neutrino transport, discussed in Chapter 11. In that problem, additional equation for the difference between the numbers of electrons and positrons per nucleon $Y_e = (n_{e^-} - n_{e^+})/n_B$ has to be solved. In that case, it is better to use as independent variables the concentrations n_ι, instead of densities $\rho_\iota = m_\iota n_\iota$.

Chapter 8

A Newton Iteration Method for Obtaining Equilibria of Rapidly Rotating Stars

A computational method for the construction of rapidly rotating stellar models is described; it combines the high (quadratic) convergence of the Newton–Raphson procedure with the use of a fine mesh, typical of self-consistent field schemes. For the presentation of the gravitational potential, a finite difference approximation to the Poisson equation is used, and the iterations of the potential are used to find the distribution of matter consistent with its gravity. It is necessary to solve a huge system of linear equations on every iteration step, but the matrix of the system is sparse, so that powerful sparse matrix solvers can be applied here. The high convergence rate of the iteration procedure and a robust sparse system solver warrant the high accuracy of calculations. The method is powerful enough for the description of models with a large density contrast and high $T/|W|$, where T and W are total kinetic and gravitational energies, respectively. The efficiency of the method is tested in calculations of the stationary states for polytropes for $T/|W|$ up to 0.425.

8.1 Introduction

Many numerical methods have been proposed for calculation of 2D or 3D self-gravitating equilibrium structures since the pioneer work done by Ivanova *et al.* [187]. One of the most advanced method, the *Self-Consistent Field (SCF) method* developed by Ostriker *et al.* [250], adopted an integral representation of the gravitational potential and used simple iterations for obtaining the equilibrium. Many useful modifications of this method have been published, e.g., [74,106,134], and the SCF method has been applied to many problems (e.g., [69,79,80,251]). An interesting version of this method has been provided by Hachisu [150]; it makes it possible to obtain solutions with high $T/|W| > 0.25$ (here T and W are total kinetic and gravitational energies of a star, respectively) or in the case of high-density contrast. By way of this method, referred to as HSCF, some 2D and 3D single- and multibody models have been calculated [125,126,150–153]. Nevertheless, both SCF and HSCF have slow (linear) convergence levels in general.

The fast (quadratic) convergence of the Newton–Raphson method constitutes another class of self-consistent techniques. In this class, the author [286] was the first to develop a method which used "accurate" formulae and a few (i.e., only six in the original work) points for representing the difference equations in the direction of the polar angle, θ, and many points and simple formulae along the radius r. The solution was obtained using the Newton–Raphson iteration procedure for the potential. On the one hand, the use of a rough grid in θ-direction significantly reduces the rank of the linear system obtained, but on the other hand, the method fails in the case of highly deformed non-spherical structures.

The authors [123] have proposed another universal and powerful method using the Newton–Raphson iteration procedure, which allows the calculation of highly deformed configurations. For the iterations, density values in grid points are used. It requires the solution of a system of linear equations with a very huge matrix whose rank is equal to the number of grid points but, contrary to the method discussed in this chapter, is dense and not sparse. The really adopted 2D grid was only 40×15 even when using a supercomputer (44×16 or 64×12 in other applications), see Refs. [124,230]. Hence, even in the case of 2D problems, this technique does not allow us to obtain a high-resolution solution, although the convergence of the method is good.

Recently, considerable progress has been attained in the numerical solution of linear systems of equations with sparse matrices of coefficients for a huge rank (some tens of thousands). This makes it possible to revert to the Newton–Raphson approach to the problem under discussion. If the Poisson equation is approximated by finite differences on the grid (instead of using the integral representation of potential) and an appropriate iteration procedure is used, then it is possible to reduce the problem to that of solving a set of linear equations with a sparse matrix. In this chapter, we suggest such a method.

Our approach is very similar to the Newton method developed in Ref. [106], who used a finite-difference approximation for the Poisson equation and considered both SCF and Newton's iterations, although his technique could be applied only to the situations where the sparse matrix had a special structure. The number of mesh points for the θ coordinate was small, and the method failed for $T/|W| \geq 0.25$. Our method allows the use of the matrix with any pattern (only its sparsity is important with respect to efficiency) so that it can be generalized to solve various physical problems. At the same time, it is also close to the HSCF method but has a higher convergence rate and it is not restricted by the number of spherical harmonics in the expansion of the potential. It allows us to obtain axisymmetric solutions with a high level of accuracy, even on a scalar computer.

Our work is a part of the project exploring the scenario for the supernova explosion mechanism suggested in Ref. [176]. It requires us to carry out 2D and 3D hydrodynamic calculations and to test the stability of very rapidly rotating models in this scenario [24]. Our technique allows us to construct very rapidly rotating configurations almost up to the limit $T/|W| = 0.5$; moreover, it is generalizable for 3D problems. The limiting configurations are also needed to obtain the integration constant for an accretion disk [67].

8.2 Basic equations

We consider stationary equilibria of axially symmetric rotating self-gravitational gases:

$$\rho(\boldsymbol{v}\nabla)\boldsymbol{v} + \nabla P + \rho\nabla\Phi = 0, \tag{8.1}$$

where ρ is the density, P is the pressure, $\boldsymbol{v}$ is the velocity of matter, and the gravitational potential Φ satisfies the Poisson equation

$$\nabla^2\Phi = 4\pi G\rho. \tag{8.2}$$

Let (r, θ, φ) be the standard spherical coordinates and (ϖ, z, φ) the cylindrical coordinates. We assume the fluid velocity without meridional circulation

$$\boldsymbol{v} = \varpi\Omega(\varpi, z)\boldsymbol{e}_{\varphi} \tag{8.3}$$

(here Ω is the angular velocity) and a *barotropic equation of state*

$$P = P(\rho). \tag{8.4}$$

In this case, it is easy to prove the *Poincaré theorem* (see Ref. [291]) on the constancy of the angular velocity on coaxial cylinders:

$$\partial\Omega(\varpi, z)/\partial z = 0. \tag{8.5}$$

Then one can obtain the Bernoulli integral of Eq. (8.1)

$$H(\rho) + \Phi + \Psi = C, \tag{8.6}$$

where C is a constant, while the *enthalpy*

$$H(\rho) = \int^{P(\rho)} \frac{dP}{\rho} \tag{8.7}$$

and the *centrifugal potential*

$$\Psi = -\int^{\varpi} \Omega^2(\varpi)\varpi d\varpi. \tag{8.8}$$

8.3 Numerical method

8.3.1 *Iteration procedure*

The choice of the basis of dimensionless variables is important for the numerical scheme. To calculate real stellar models, when the total mass M must remain constant along a rotating sequence, M shall be specified as one of the basic dimensional units. In this case, our method has to use a relation similar to Eq. (31) in Ref. [106], which is easily incorporated in our scheme. For the barotropic sequences, a simpler approach is possible when we adopt the following basic dimension units: G — the gravitational constant, R_e — the equatorial radius of the star, and ρ_0 — the value of density in a fixed point labeled by the subscript 0. Then the unit for Φ, Ψ, H is $\Phi_u = GR_e^2\rho_0$. One can express the pressure as $P = GR_e^2\rho_0^2\hat{P}$, the square of the angular velocity as $\Omega^2 = G\rho_0\hat{\Omega}^2$, the star mass as $M = R_e^3\rho_0\hat{M}$, the angular momentum as $J = G^{1/2}R_e^5\rho_0^{3/2}\hat{J}$, the gravitational energy as $W = GR_e^5\rho_0^2\hat{W}$, and the kinetic energy as $T = GR_e^5\rho_0^2\hat{T}$. Here we denote as $\hat{f}$ the dimensionless form of a variable f. We have to solve the following system of equations (the operator ∇^2 is assumed in dimensionless variables):

$$\rho_0\hat{\rho} = H^{-1}[\Phi_u(\hat{C} - \hat{\Phi} - \hat{\Psi})], \tag{8.9}$$

$$\nabla^2\hat{\Phi} = 4\pi\hat{\rho}. \tag{8.10}$$

In the point 0: $\rho = \rho_0$,

$$\Phi_u(\hat{C} - \hat{\Phi}_0 - \hat{\Psi}_0) = H(\rho_0), \tag{8.11}$$

and in the two fixed boundary points A, B where $\rho = 0$ [150]:

$$\Phi_u(\hat{C} - \hat{\Phi}_A - \hat{\Psi}_A) = H(0), \tag{8.12}$$

$$\Phi_u(\hat{C} - \hat{\Phi}_B - \hat{\Psi}_B) = H(0). \tag{8.13}$$

Of course, one can choose the definition of enthalpy in (8.7) in the most natural way putting $H(0) = 0$, but we retain $H(0)$ in the following for the sake of generality. The point A will be chosen as a point on the star surface in the equatorial plane so that $\rho(r > r_A) = 0$.

It is convenient to take the function $\hat{\Psi}$ in the form

$$\hat{\Psi} = C_\Psi\hat{\Psi}^0(\hat{\varpi}), \tag{8.14}$$

where C_Ψ is an unknown constant and $\hat{\Psi}^0(\hat{\varpi})$ is a prescribed function, for example [123],

$$\hat{\Psi}^0(\hat{\varpi}) = \begin{cases} -\hat{\varpi}^2/2 & \textit{rigid rotation}, \\ -\frac{1}{2}\ln(d^2 + \hat{\varpi}^2) & v\textit{-constant rotation}, \\ -1/[2(d^2 + \hat{\varpi}^2)] & j\textit{-constant rotation}, \end{cases} \tag{8.15}$$

where d is some constant.

From conditions (8.11), (8.12), and (8.13), it is possible to express the constants C_Ψ, $\hat{C}$, $\hat{\Phi}_0$ through the values of $\hat{\Phi}$ in points 0, A, B:

$$C_\Psi = \frac{\hat{\Phi}_B - \hat{\Phi}_A}{\hat{\Psi}^0_A - \hat{\Psi}^0_B}, \tag{8.16}$$

$$\hat{C} = \frac{\hat{\Phi}_A + \hat{\Psi}_A - (\hat{\Phi}_0 + \hat{\Psi}_0)H(0)/H(\rho_0)}{1 - H(0)/H(\rho_0)}, \tag{8.17}$$

$$\Phi_u = \frac{H(\rho_0)}{\hat{C} - \hat{\Phi}_0 - \hat{\Psi}_0}, \tag{8.18}$$

and rewrite Eq. (8.9)

$$\rho_0 \hat{\rho} = H^{-1}\left[\frac{H(\rho_0)}{\hat{C} - \hat{\Phi}_0 - \hat{\Psi}_0}(\hat{C} - \hat{\Phi} - \hat{\Psi})\right], \tag{8.19}$$

where $\hat{\Psi} = C_\Psi \hat{\Psi}^0$, and C_Ψ, $\hat{C}$ are functions of the unknown function $\hat{\Phi}$. So, we have to define $\hat{\Phi}$ from (8.10) and (8.19).

We shall use the *Newton iteration procedure* for this, and for the $(k+1)$th iteration step Eqs. (8.10) and (8.19) give the following:

$$\begin{aligned} \nabla^2 \hat{\Phi}^{k+1} = 4\pi \Bigg\{ \rho^k + \frac{(H^{-1})'}{\rho_0} \frac{H(\rho_0)}{\hat{C} - \hat{\Phi}_0 - \hat{\Psi}_0} \Bigg[&(\hat{C}^{k+1} - \hat{C}^k) \\ &- (\hat{\Phi}^{k+1} - \hat{\Phi}^k) - (\hat{\Psi}^{k+1} - \hat{\Psi}^k) - \Big((\hat{C}^{k+1} - \hat{C}^k) \\ &- (\hat{\Phi}^{k+1}_0 - \hat{\Phi}^k_0) - (\hat{\Psi}^{k+1}_0 - \hat{\Psi}^k_0)\Big) \frac{\hat{C} - \hat{\Phi}_0 - \hat{\Psi}_0}{\hat{C} - \hat{\Phi} - \hat{\Psi}} \Bigg] \Bigg\}, \end{aligned} \tag{8.20}$$

where the quantities without superscripts are taken at the kth step of iteration, and $(\hat{C}^{k+1} - \hat{C}^k)$, $(\hat{\Psi}^{k+1} - \hat{\Psi}^k)$, and $(\hat{\Psi}^{k+1}_0 - \hat{\Psi}^k_0)$ should be expressed through $(\hat{\Phi}^{k+1}_0 - \hat{\Phi}^k_0)$, $(\hat{\Phi}^{k+1}_A - \hat{\Phi}^k_A)$, and $(\hat{\Phi}^{k+1}_B - \hat{\Phi}^k_B)$ from (8.14), (8.16), and (8.17). So, actually, the unknown function is $\hat{\Phi}$.

In the general barotropic case, the order of calculations should be the following: to set ρ_0, in using the iteration procedure (8.20) to find $\hat{\Phi}$, and then to find R_e from the relation $GR_e^2\rho_0 = \Phi_u$ (Φ_u is found from (8.18)).

In the polytropic case: $P = K\rho^{1+1/n}$, $H(\rho) = (1+n)K\rho^{1/n}$, there is a free scaling parameter; for our choice of units, it is ρ_0. It is clear that ρ_0 disappears in Eq. (8.19) so that we can find the set of solutions for different ρ_0 by solving one problem with fixed points 0, A, B.

8.3.2 *Computational grid*

There are two general ways to obtain a finite representation of continuous fields (like Φ and ρ). One way is to use base functions and truncated series for the fields expanded in this basis (like Ref. [250]). Another way is to use a direct finite difference representation (like Ref. [106]). It is common practice to combine finite differences for r and expansions in Legendre polynomials for θ. The basis function approach has many advantages (see Ref. [160] for the discussion of this approach in the context of stellar dynamics). We have gained some experience using this approach in the SCF method: The use of fundamental solutions by way of the Laplace equation in spheroidal coordinates allows us to obtain very flattened configurations [70]. In this work, we want to develop a robust method that produces an equilibrium solution in the same representation as used in multidimensional hydrodynamic calculations, that is, in the finite-difference one.

For the solution of Eq. (8.20), we follow some ideas developed in Ref. [106], i.e., we introduce a uniform grid for the radial coordinate inside a sphere, encircling the bulk of the star mass, and we use an inversion to solve the Laplace equation up to infinite r. We use the grid $(\hat{r}_i, \theta_j)$ in the spherical coordinate system. The r-mesh coordinate is taken to be

$$\hat{r}_i = \frac{i}{i_A}, \; \Delta\hat{r}_i \equiv \hat{r}_i - \hat{r}_{i-1} = \text{constant}, \; \text{at } 0 \le i \le i_A, \tag{8.21}$$

where i_A is the point on the star surface in the equatorial plane, and

$$\frac{1}{\hat{r}_{i-1}} - \frac{1}{\hat{r}_i} \equiv h = \text{constant}, \; \text{at } i_A \le i \le i_{\max}, \tag{8.22}$$

$\hat{r}_{i_{\max}} = \infty$. So,

$$h = \frac{1}{\hat{r}_{i_A-1}} - \frac{1}{\hat{r}_{i_A}} = \frac{1}{i_A - 1} = \frac{1}{\hat{r}_{i_{\max}-1}} = \frac{1}{i_{\max} - i_A}, \tag{8.23}$$

and

$$i_A = (i_{\max} + 1)/2 \tag{8.24}$$

should be taken.

The θ-mesh coordinate can be chosen in the following form (the plane $\theta = \pi/2$ is the plane of symmetry):

$$\theta_j = \frac{\pi}{2}\frac{j}{j_{\max}}, \quad 0 \le j \le j_{\max}. \tag{8.25}$$

The radial part of operator ∇^2:

$$\nabla_r^2\hat{\Phi} \equiv \frac{1}{\hat{r}^2}\frac{\partial}{\partial \hat{r}}\left(\hat{r}^2\frac{\partial\hat{\Phi}}{\partial\hat{r}}\right) = \frac{1}{\hat{r}^4}\frac{\partial^2\hat{\Phi}}{\partial(1/\hat{r})^2}. \tag{8.26}$$

In the region $\hat{r} < 1$ $(i \le i_A - 1)$, we use the following approximation of ∇_r^2:

$$\nabla_r^2\hat{\Phi}(\hat{r}_i) \simeq \frac{1}{(\Delta\hat{r})^2\hat{r}_i^2}\left[\frac{\hat{r}_{i+1}^2 + \hat{r}_i^2}{2}(\hat{\Phi}_{i+1} - \hat{\Phi}_i) - \frac{\hat{r}_i^2 + \hat{r}_{i-1}^2}{2}(\hat{\Phi}_i - \hat{\Phi}_{i-1})\right], \tag{8.27}$$

and in the region $1 \le \hat{r}$ $(i_A \le i \le i_{\max})$,

$$\nabla_r^2\hat{\Phi}(\hat{r}_i) \simeq \frac{1}{\hat{r}_i^4 h^2}\left[\hat{\Phi}_{i-1} - 2\hat{\Phi}_i + \hat{\Phi}_{i+1}\right]. \tag{8.28}$$

The angular part of ∇^2:

$$\nabla_\theta^2\hat{\Phi} \equiv \frac{1}{\hat{r}^2\sin\theta}\frac{\partial}{\partial\theta}\left(\sin\theta\frac{\partial\hat{\Phi}}{\partial\theta}\right) \tag{8.29}$$

is approximated by

$$\nabla_\theta^2\hat{\Phi}(\hat{r}_i, \theta_j) \simeq \frac{1}{\hat{r}_i^2\sin\theta_j(\Delta\theta)^2}\left[\frac{\sin\theta_{j+1} + \sin\theta_j}{2}(\hat{\Phi}_{i,j+1} - \hat{\Phi}_{i,j}) - \frac{\sin\theta_j + \sin\theta_{j-1}}{2}(\hat{\Phi}_{i,j} - \hat{\Phi}_{i,j-1})\right]. \tag{8.30}$$

Now we describe our treatment of the boundary conditions. For $\hat{r} \to 0$, we have

$$\int_0^\pi \nabla^2\hat{\Phi}\sin\theta d\theta \to 3\int\frac{\partial^2\hat{\Phi}}{\partial\hat{r}^2}\sin\theta d\theta, \tag{8.31}$$

so

$$4\pi\hat{\rho}_{0,k} = \int_0^\pi \nabla^2\hat{\Phi}\sin\theta d\theta \mid_{\hat{r}=0}$$
$$\simeq \frac{6}{(\Delta\hat{r})^2}\left(\sum_k \{\cos[\max(0,\Delta\theta(k-1/2))]\right.$$
$$\left. - \cos[\min(\pi/2,\Delta\theta(k+1/2))]\}(\hat{\Phi}_{1,k}-\hat{\Phi}_{0,j})\right), \tag{8.32}$$

where we use the notation $\hat{\Phi}_{i,j} \equiv \hat{\Phi}(\hat{r}_i,\theta_j)$. For $\hat{r}\to\infty$, we have $\hat{\Phi}=0$:

$$\hat{\Phi}_{i_{\max},j} = 0. \tag{8.33}$$

At $\theta = 0$,

$$\nabla_\theta^2\hat{\Phi}_{i,0} = \frac{2}{\hat{r}^2}\frac{\partial^2\hat{\Phi}}{\partial\theta^2}\mid_{i,0}\simeq \frac{2}{\hat{r}_i^2}\frac{2\hat{\Phi}_{i,1}-2\hat{\Phi}_{i,0}}{(\Delta\theta)^2}. \tag{8.34}$$

At $\theta = \pi/2$,

$$\nabla_\theta^2\hat{\Phi}_{i,j_{\max}} = \frac{1}{\hat{r}^2}\frac{\partial^2\hat{\Phi}}{\partial\theta^2}\mid_{i,j_{\max}}\simeq \frac{1}{\hat{r}_i^2}\frac{2\hat{\Phi}_{i,j_{\max}-1}-2\hat{\Phi}_{i,j_{\max}}}{(\Delta\theta)^2}. \tag{8.35}$$

From (8.20), we have the system of linear equations for finding $\hat{\Phi}_{i,j}$. If we denote $u_k = \hat{\Phi}_{i,j}$ where $k = k(i,j) = 1 + i(j_{\max}+1) + j$, then we have the sparse system

$$A\boldsymbol{u} = \boldsymbol{f} \tag{8.36}$$

with non-zero elements:

$a_{i,i},\ a_{i-1,i},\ a_{i+1,i},\ a_{i-(j_{\max}+1),i},\ a_{i+(j_{\max}+1),i},\ a_{i,k_{i_0,j_0}},\ a_{i,k_{0,0}},\ a_{i,k_{i_A,j_A}},$
$a_{i,k_{i_B,j_B}},\ a_{k_{0,j_{\max}},k_{1,0}\le\alpha\le k_{1,j_{\max}-1}},$

where $1 \le i \le k_{\max} = k_{i_{\max},j_{\max}} = (i_{\max}+1)(j_{\max}+1)$.

For the solution of the sparse system (8.36), we use the elimination technique developed by Zlatev [248]. We use our code designed according to their specifications: The choice of the pivots in *LU* decomposition follows the Zlatev strategy, and the iterative refinement may be used for the improvement of the solution. We find that the iterative refinement is not needed for the value of the Zlatev threshold $\leq 10^{-8}$. For the threshold $\approx 10^{-4}$, the iterative refinement is necessary, but then the LU-decomposed matrix is not so dense as for smaller thresholds, and the algorithm is $\approx 30\%$ faster.

The elements of the matrix A in (8.36) change on every iteration step, but the structure of A (the position of non-zero elements) remains the same. This fact is taken into account by the linear equation solver used.

8.3.3 *Numerical example: Rotating polytropes*

In our example, *rigidly rotating polytropes with* $n = 1.5$ with spheroidal structure were calculated. We adopted $r_0 = 0$; $r_A = 1$, $\theta_A = \pi/2$, and the boundary point B was taken on the polar axis: $r_B \leq 1$, $\theta_B = 0$ (following Ref. [150]).

The calculations started from a spherical configuration: $r_B = 1$. After calculating one model, we reduced r_B and calculated the new model with $\hat{\Phi}$ from the preceding step as an initial approximation for $\hat{\Phi}$. The initial approximation for the spherical configuration was rather arbitrary, for example,

$$\hat{\Phi} = \begin{cases} -1.5, & \hat{r} < 1, \\ -1, & \hat{r} = 1, \\ -0.5, & \hat{r} > 1. \end{cases} \tag{8.37}$$

The iterations were continued until the correction for C was small: $\delta\hat{C} < 10^{-7}$. The number of iterations depended, of course, on the proximity of the initial approximation to the final solution. In the following tables, we show all actual sequences so that initial approximations appear rather crude. Nevertheless, the number of iterations (≤ 4) was appreciably lower than in SCF methods, especially for very high T/W.

In Table 8.1, we present the dimensionless values for $\hat{r}_B$, C_Ψ, and for the following integrals:

$$\hat{M} = \sum \hat{\rho}_{ij}\Delta V_{ij}, \tag{8.38}$$

$$\hat{V} = \sum_{\hat{\rho}_{ij}>0} \Delta V_{ij}, \tag{8.39}$$

$$\hat{J} = \sum \hat{\rho}_{ij}\hat{v}_{\phi_{ij}}\hat{r}_i \sin\theta_j \Delta V_{ij}, \tag{8.40}$$

Table 8.1 Polytropes $n = 1.5$ (rigid rotation, spheroidal configurations) [12].

$\hat{r}_B$	C_Ψ	$\hat{M}$	$\hat{V}$	$\hat{J}$	$-\hat{W}$	$\hat{P}_{\max}$	$\hat{T}/(-\hat{W})$	VT
1.000	0.0000	0.6993	4.16	0.00000	0.41920	0.3765	0.00000	$9.4 \cdot 10^{-6}$
0.900	0.1033	0.5974	3.76	0.03762	0.32070	0.3276	0.01885	$7.1 \cdot 10^{-6}$
0.800	0.1913	0.4883	3.28	0.03905	0.22750	0.2765	0.03754	$9.3 \cdot 10^{-6}$
0.750	0.2266	0.4304	3.03	0.03557	0.18350	0.2499	0.04614	$9.6 \cdot 10^{-6}$
0.700	0.2539	0.3700	2.75	0.03013	0.14210	0.2227	0.05343	$9.6 \cdot 10^{-6}$
0.650	0.2705	0.3072	2.41	0.02330	0.10390	0.1949	0.05832	$9.6 \cdot 10^{-6}$
0.625	0.2738	0.2752	2.21	0.01961	0.08643	0.1807	0.05937	$9.1 \cdot 10^{-6}$

$$-\hat{W} = -\frac{1}{2}\sum \hat{\rho}_{ij}\hat{\Phi}_{ij}\Delta V_{ij}, \tag{8.41}$$

$$\hat{T} = \frac{1}{2}\sum \hat{\rho}_{ij}\hat{v}^2_{\phi_{ij}}\Delta V_{ij}, \tag{8.42}$$

where

$$\Delta V_{ij} = 4\pi \frac{\left(\hat{r}_i + \hat{r}_{\min(i+1,i_{\max})}\right)^3 - \left(\hat{r}_{\max(i-1,0)} + \hat{r}_i\right)^3}{3 \times 2^3}$$
$$\times \left(\cos\frac{\theta_{\max(j-1,0)} + \theta_j}{2} - \cos\frac{\theta_j + \theta_{\min(j-1,j_{\max})}}{2}\right), \tag{8.43}$$

$$\hat{v}_\phi = \hat{\Omega}\hat{\varpi} = \left(-\frac{\partial\hat{\Psi}}{\partial\hat{\varpi}}\hat{\varpi}\right)^{1/2}, \tag{8.44}$$

and for the quantity:

$$\hat{P}_{\max} = \max_{ij}\hat{P}_{ij}. \tag{8.45}$$

The *virial test*

$$VT = \frac{|2\hat{T} + \hat{W} + 3\hat{\Pi}|}{|\hat{W}|}, \tag{8.46}$$

where

$$\hat{\Pi} = \sum \hat{P}_{ij}\Delta V_{ij}, \tag{8.47}$$

shows the accuracy of the numerical solution. On the computational grid used, $i_{\max} = 401$, $j_{\max} = 50$, we have $VT < 10^{-5}$. Our results (Table 8.1) are very close to those obtained by [150].

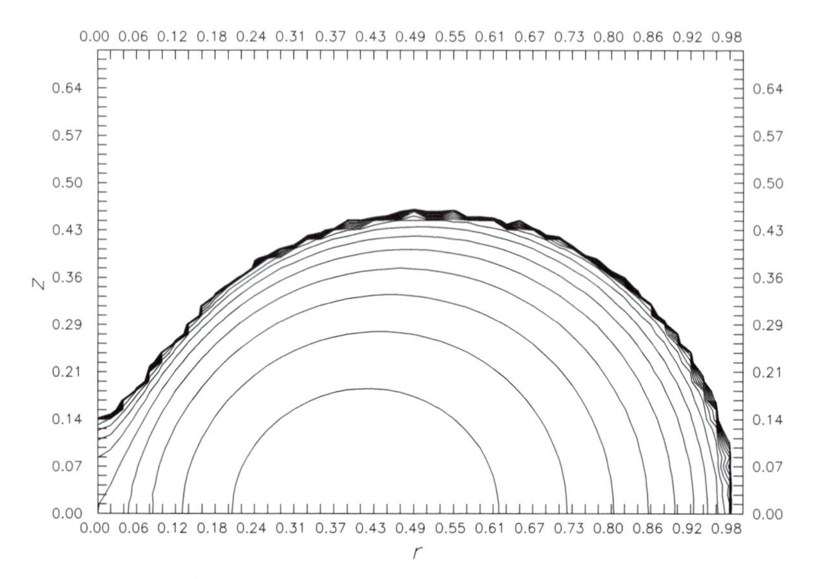

Figure 8.1 The density contours for the v-constant law for polytrope $n = 1.5$ and axis ratio $r_B = 0.167$ ($\log_{10} \rho_{min} = -3$, $\log_{10} \rho_{max} = 0$, $\Delta \log_{10} \rho = 0.3$) [12].

We ran the polytrope $n = 1.5$ for the sequence with differential rotation (the case of v-constant rotation in 8.15). In Fig. 8.1, we present the isopycnic contours for one of the models in this sequence with its "dumbbell" shape for the axis ratio 0.167.

One can compare our results for this model with those of Ref. [152] in Table 2. Hachisu used a somewhat different system of units, based on the maximum value of density $\rho_{\max}$, instead of our ρ_0. For the model with $r_B = 0.167$, we obtain the following results in the Hachisu units: $M = 0.6129$, $V = 2.321$, $J = 1.506$, $T = 0.06928$, $W = -0.3341$, and $P_{\max} = 0.1652$. The virial test for this model is $VT = 3.1 \cdot 10^{-5}$.

The method also allows one to obtain the toroidal structures (we have done this but do not present the results, which are again close to Ref. [150]). For this, it is necessary to set the point B on the equatorial plane, to select the point 0 inside the star (e.g., between points A and B), and to take an appropriate initial approximation for $\hat{\Phi}$.

8.3.3.1 *Rapidly rotating polytropes: self-gravitating thick and slim disks*

The described method allows us to construct the rapidly rotation models very close to the limit

$$T/|W| = 0.5, \tag{8.48}$$

which follows from the virial theorem $2T + W + 3\Pi = 0$. In order to construct a sequence of very flattened models, we suggest that the rotation law be defined in such a form that the centrifugal force forms certain fraction α of the gravitational force at the equatorial plane ($\theta = \pi/2$):

$$\alpha \frac{\partial \hat{\Phi}}{\partial \hat{r}} = \hat{\Omega}^2 \hat{r} = -\frac{\partial \hat{\Psi}}{\partial \hat{r}}, \tag{8.49}$$

where $0 \leq \alpha = \text{constant} < 1$. Let us refer to this as the α-constant rotation law. We can take the centrifugal potential $\hat{\Psi}(\hat{\varpi})$ from (8.14) in the following form:

$$\hat{\Psi}(\hat{\varpi}) = -\alpha \begin{cases} \hat{\Phi}(\hat{r} = \hat{\varpi}, \pi/2) - \hat{\Phi}_0, & \hat{\varpi} < 1, \\ \hat{\Phi}(1, \pi/2) - \hat{\Phi}_0, & \hat{\varpi} > 1. \end{cases} \tag{8.50}$$

For spheroidal configurations (and for the same fixed points 0, A, B as in the preceding section), the conditions (8.11), (8.12), and (8.13) give

$$\hat{C} = \frac{\hat{\Phi}_B - H(0)/H(\rho_0)\hat{\Phi}_0}{1 - H(0)/H(\rho_0)}, \tag{8.51}$$

$$\Phi_u = \frac{H(\rho_0)}{\hat{C} - \hat{\Phi}_0}, \tag{8.52}$$

and

$$\hat{\Psi}(\hat{\varpi}) = \begin{cases} \frac{\hat{\Phi}_B - \hat{\Phi}_A}{\hat{\Phi}_A - \hat{\Phi}_0} (\hat{\Phi}(\hat{r} = \hat{\varpi}, \pi/2) - \hat{\Phi}_0), & \hat{\varpi} < 1, \\ \hat{\Phi}_B - \hat{\Phi}_A, & \hat{\varpi} > 1. \end{cases} \tag{8.53}$$

In the main iteration relation (8.20), we now have more the complicated expression for the difference $(\hat{\Psi}^{k+1} - \hat{\Psi}^k)$ through the values of Φ.

We find

$$(\hat{\Psi}^{k+1} - \hat{\Psi}^{k}) = \frac{\partial \hat{\Psi}}{\partial \hat{\Phi}_0}\left(\hat{\Phi}_0^{k+1} - \hat{\Phi}_0^{k}\right) + \frac{\partial \hat{\Psi}}{\partial \hat{\Phi}_A}\left(\hat{\Phi}_A^{k+1} - \hat{\Phi}_A^{k}\right) + \frac{\partial \hat{\Psi}}{\partial \hat{\Phi}_B}\left(\hat{\Phi}_B^{k+1} - \hat{\Phi}_B^{k}\right) + \frac{\partial \hat{\Psi}}{\partial \hat{\Phi}\left(\hat{\varpi}, \frac{\pi}{2}\right)} \times \left(\hat{\Phi}^{k+1}\left(\hat{\varpi}, \frac{\pi}{2}\right) - \hat{\Phi}^{k}\left(\hat{\varpi}, \frac{\pi}{2}\right)\right), \quad (8.54)$$

and $\hat{\Psi}_0 \equiv 0$.

In the computational grid, we define $\hat{\varpi}_{ij} = \hat{r}_i \sin\theta_j$, and we use the following approximation:

$$\hat{\Phi}(\hat{r} = \hat{\varpi}_{ij}, \theta = \pi/2) \simeq c_1 \hat{\Phi}_{i_e - 1, j_{\max}} + c_2 \hat{\Phi}_{i_e, j_{\max}}, \quad (8.55)$$

where i_e is the nearest to $\hat{\varpi}_{ij}$ point such that $\hat{r}_{i_e-1} \le \hat{\varpi}_{ij} \le \hat{r}_{i_e}$, and $c_1 = (\hat{r}_{i_e} - \hat{\varpi}_{ij})/(\hat{r}_{i_e} - \hat{r}_{i_e-1})$, $c_2 = (\hat{\varpi}_{ij} - \hat{r}_{i_e-1})/(\hat{r}_{i_e} - \hat{r}_{i_e-1})$. Hence, in the matrix A (8.36), new non-zero elements $a_{k_{i,j}, k_{i_e(i,j)-1, j_{\max}}}$, $a_{k_{i,j}, k_{i_e(i,j), j_{\max}}}$, $0 \le i \le i_{\max}$, $0 \le j \le j_{\max}$ appear.

In Tables 8.2 and 8.3, we present our results obtained for the α-constant law for the polytropic indexes $n = 1.5$ and $n = 3$ on the grid 200×51. The notation is the same as for Table 8.1. Figure 8.2

Table 8.2 Rapidly rotating polytropes $n = 1.5$ (spheroidal configurations) for the α-constant law [12].

$\hat{r}_B$	α	$\hat{M}$	$\hat{V}$	$\hat{J}$	$-\hat{W}$	$\hat{P}_{\max}$	$\hat{T}/(-\hat{W})$	VT
1.00	0.000	0.7000	4.130	0.00000	0.42000	0.37700	0.0000	$3.8 \cdot 10^{-5}$
0.80	0.120	0.5790	3.540	0.05810	0.30400	0.30000	0.0483	$2.8 \cdot 10^{-5}$
0.70	0.187	0.5160	3.200	0.06290	0.24900	0.25900	0.0772	$2.5 \cdot 10^{-5}$
0.60	0.262	0.4500	2.860	0.06250	0.19700	0.21700	0.1100	$2.0 \cdot 10^{-5}$
0.50	0.344	0.3830	2.480	0.05810	0.14800	0.17400	0.1490	$1.5 \cdot 10^{-5}$
0.40	0.436	0.3150	2.110	0.05040	0.10400	0.13000	0.1940	$8.1 \cdot 10^{-6}$
0.30	0.542	0.2440	1.680	0.03990	0.06480	0.08710	0.2480	$5.0 \cdot 10^{-6}$
0.25	0.601	0.2070	1.460	0.03360	0.04790	0.06650	0.2790	$1.5 \cdot 10^{-5}$
0.20	0.666	0.1690	1.220	0.02690	0.03280	0.04710	0.3140	$3.4 \cdot 10^{-5}$
0.15	0.737	0.1300	0.963	0.01960	0.01990	0.02950	0.3530	$6.7 \cdot 10^{-5}$
0.12	0.783	0.1060	0.800	0.01510	0.01340	0.02010	0.3790	$1.1 \cdot 10^{-4}$
0.10	0.816	0.0897	0.689	0.01210	0.00963	0.01460	0.3970	$1.6 \cdot 10^{-4}$
0.09	0.832	0.0812	0.625	0.01060	0.00794	0.01210	0.4060	$2.0 \cdot 10^{-4}$
0.08	0.849	0.0726	0.569	0.00911	0.00638	0.00979	0.4160	$2.9 \cdot 10^{-4}$
0.07	0.867	0.0639	0.499	0.00764	0.00496	0.00766	0.4250	$4.4 \cdot 10^{-4}$

Table 8.3 Rapidly rotating polytropes $n = 3$ (spheroidal configurations) for the α-constant law [12].

$\hat{r}_B$	α	$\hat{M}$	$\hat{V}$	$\hat{J}$	$-\hat{W}$	$\hat{P}_{\max}$	$\hat{T}/(-\hat{W})$	VT
1.00	0.0000	0.0775	4.13	0.00000	0.00901	0.0662	0.0000	$5.0 \cdot 10^{-5}$
0.80	0.0639	0.0661	3.81	0.00127	0.00686	0.0562	0.0272	$6.0 \cdot 10^{-5}$
0.70	0.1040	0.0596	3.61	0.00140	0.00574	0.0504	0.0447	$6.9 \cdot 10^{-5}$
0.60	0.1510	0.0526	3.38	0.00142	0.00463	0.0441	0.0658	$8.3 \cdot 10^{-5}$
0.50	0.2070	0.0451	3.13	0.00134	0.00354	0.0372	0.0919	$1.1 \cdot 10^{-4}$
0.40	0.2770	0.0370	2.84	0.00118	0.00250	0.0297	0.1250	$1.5 \cdot 10^{-4}$
0.30	0.3660	0.0286	2.50	0.00094	0.00158	0.0216	0.1690	$2.2 \cdot 10^{-4}$
0.25	0.4220	0.0243	2.30	0.00080	0.00118	0.0175	0.1970	$2.9 \cdot 10^{-4}$
0.20	0.4890	0.0199	2.08	0.00066	0.00082	0.0133	0.2310	$3.9 \cdot 10^{-4}$
0.15	0.5700	0.0156	1.80	0.00050	0.00052	0.0091	0.2740	$5.5 \cdot 10^{-4}$
0.12	0.6300	0.0129	1.61	0.00040	0.00036	0.0067	0.3060	$7.0 \cdot 10^{-4}$
0.10	0.6760	0.0111	1.46	0.00034	0.00027	0.0051	0.3300	$8.4 \cdot 10^{-4}$
0.09	0.7000	0.0102	1.36	0.00031	0.00023	0.0044	0.3440	$9.2 \cdot 10^{-4}$
0.08	0.7270	0.0093	1.28	0.00028	0.00019	0.0036	0.3580	$1.0 \cdot 10^{-3}$
0.07	0.7550	0.0084	1.19	0.00024	0.00016	0.0029	0.3703	$1.1 \cdot 10^{-3}$

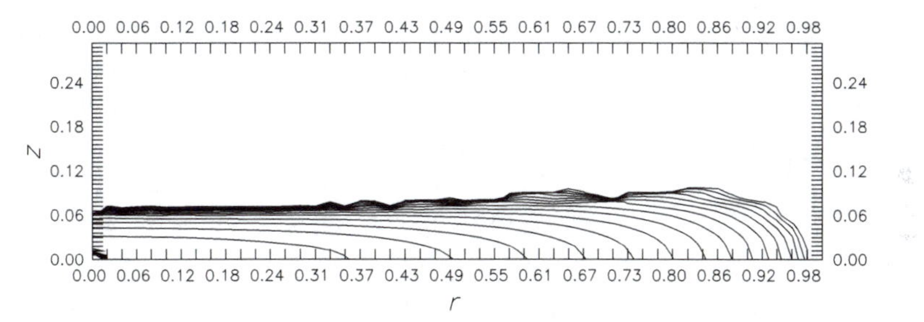

Figure 8.2 The density contours for the α-constant law for polytrope $n = 1.5$ and axis ratio $r_B = 0.07$ ($\log_{10} \rho_{\min} = -3$, $\log_{10} \rho_{\max} = 0$, $\Delta \log_{10} \rho = 0.2$) [12].

shows the contours of isopycnic surfaces for the model with axis ratio $r_B = 0.07$ in this sequence. Unlike in the case of v- or j-constant laws, the shape of the configuration for the α-constant law is much flatter.

8.4 Discussion

The method has the second-order accuracy (relative to the size of the space zones). The dependence of solution accuracy ($-\hat{C}$, VT) on

Table 8.4 Dependence of the algorithm accuracy on the number of grid points (1D case) [12].

i_A	$-\hat{C}$	VT
10	0.705	$5.6 \cdot 10^{-3}$
20	0.703	$1.2 \cdot 10^{-3}$
50	0.7002	$1.6 \cdot 10^{-4}$
100	0.69954	$3.8 \cdot 10^{-5}$
200	0.69932	$9.1 \cdot 10^{-6}$
500	0.69924	$1.4 \cdot 10^{-6}$
1000	0.699221	$3.5 \cdot 10^{-7}$
2000	0.699217	$8.6 \cdot 10^{-8}$
5000	0.6992153	$1.4 \cdot 10^{-8}$

the number of zones (i.e., on $i_{\max}$) is presented in Table 8.4 for the 1D case (i.e., without rotation) for $n = 1.5$.

One can see that $VT \propto i_{\max}^{-2} \propto (\Delta\hat{r})^2$, where $\Delta\hat{r}$ is the radial zone size. Thus, for solutions deviating considerably from the spherical shape, the grid should satisfy the following condition: $\hat{r}_i \Delta\theta_j \simeq \Delta\hat{r}_i$.

The number of iterations is virtually independent of the grid size, and the CPU time spent on one iteration is $\propto N_z^\beta$, where N_z is the number of non-zero elements in the matrix A from (8.36) and $\beta < 2$.

Though our method is faster than the Newton iteration method developed in Ref. [123], it is still rather slow on large grids (see details in Ref. [12]). One way to speed up the algorithm would be to use a vectorized version of our sparse matrix solver. Our experience shows that, in general, the use of special built-in routines like `SCATTER` and `GATHER` allows us to save about 50% of the CPU time on a vector computer. Another way could be to use other sparse matrix solvers, designed especially for use on the vector machines. More numerical experiments are needed here however.

In principle, it is possible to extend this method to the 3D case [150–152], but the grid (using our matrix solver and workstations of the current power) will not be very fine, e.g., $100 \times 30 \times 30$.

In conclusion, we discuss briefly the prospects of application of our method to construction of the models which are more realistic than polytropes. The application of our method is straightforward when the barotropic equation of state (8.4) is the good approximation to reality (e.g., in the case of degenerate stars). For a general equation of state $P = P(\rho, S)$, where S is the entropy, or $P = P(\rho, T_p)$, where T_p is the temperature, the rotation law (8.5) is distinguished since it is necessary for the secular stability of a rotating star [133,144]. Then the surfaces of constant pressure, density, and temperature all coincide and for this pseudobarotropic situation, our method may be generalized following the pioneering works [79,186,216] who considered the application of the SCF method to real hot stars with radiation. The authors [125,126] have developed the generalization of the HSCF method for radiating rapidly rotating pseudobarotropic stars and toroids with the account of meridional circulation. Such a generalization of our method is also possible. We note a very important paper [255] on the meridional circulation for very rapidly rotating stars: Their approach together with that of Ref. [125] may be combined with our method. This generalization, as well as application of the method to baroclinic stars [301], requires, of course, much additional work.

Chapter 9

Numerical Solution of the Poisson Equation for the 3D Modeling of Stellar Evolution

In this section, an effective algorithm for solving the Poisson equation which is used in the finite-difference scheme for the integration of hydrodynamic equations on a stationary grid in spherical coordinates is presented. The method is based on spherical functions and can be used in 1D, 2D, and 3D [10].

Then finite-difference hydrodynamic equations with self-gravitation are solved, the potential $\Phi_{i,j,k}$, and the gravitational acceleration $\boldsymbol{g}_{i,j,k} \equiv (f_{i,j,k}, g_{i,j,k}, h_{i,j,k})$, in the grid cell $(r_{i-1/2}, r_{i+1/2}) \times (\theta_{j-1/2}, \theta_{j+1/2}) \times (\phi_{k-1/2}, \phi_{k+1/2})$ with volume $\Delta V_{i,j,k}$ and mass $\Delta m_{i,j,k} = \rho_{i,j,k}\Delta V_{i,j,k}$ must be defined in each time step with a different distribution of density $\{\rho_{i,j,k}\}$. Eulerian coordinates and a stationary grid are assumed to be used in the difference scheme.

Using the definition of *Legendre polynomials in terms of the generating function*, we can rewrite the integral representation of the potential

$$-\Phi(\boldsymbol{R}) = \int d\boldsymbol{r} \frac{G\rho(\boldsymbol{r})}{|\boldsymbol{R} - \boldsymbol{r}|}$$

in the form

$$-\Phi(\boldsymbol{R}) = \int_{r<R} d\boldsymbol{r} G\rho \frac{1}{R} \sum_{l=0}^{\infty} \left(\frac{r}{R}\right)^l P_l(\cos\mu)$$

$$+ \int_{r>R} d\boldsymbol{r} G\rho \frac{1}{r} \sum_{l=0}^{\infty} \left(\frac{R}{r}\right)^l P_l(\cos\mu),$$

where

$$\cos\mu = \frac{\boldsymbol{r}\boldsymbol{R}}{rR}.$$

According to the *addition theorem for associated Legendre polynomials* [310]

$$\Phi(\boldsymbol{R}) = \sum_{l=0}^{\infty} \sum_{m=0}^{l} \left(\Phi_{lm}^{<}(\boldsymbol{R}) + \Phi_{lm}^{>}(\boldsymbol{R})\right), \tag{9.1}$$

where

$$-\Phi_{lm}^{<}(\boldsymbol{R}) = \int_{r<R} d\boldsymbol{r} G\rho \frac{1}{R} \left(\frac{r}{R}\right)^l Q_l^m(\cos\theta) Q_l^m(\cos\Theta) \cos(m(\tilde{\Phi} - \phi))$$

and

$$-\Phi_{lm}^{>}(\boldsymbol{R}) = \int_{r>R} d\boldsymbol{r} G\rho \frac{1}{r} \left(\frac{R}{r}\right)^l Q_l^m(\cos\theta) Q_l^m(\cos\Theta) \cos(m(\tilde{\Phi} - \phi)),$$

$$Q_l^m(\cos\theta) \equiv P_l^m(\cos\theta) \begin{cases} \sqrt{2\dfrac{(l-m)!}{(l+m)!}}, & m \neq 0, \\ 1, & m = 0. \end{cases}$$

Thus, the expansion is actually performed in terms of powers of x^l, $l \geq 0$, and $x < 1$.

Assuming that the mass is distributed uniformly over the cell volume, let us determine the field at point $\boldsymbol{R}$ produced by the cell i, j, k.

We use the following notations: the averaging over r,

$$\langle f(r)\rangle_i \equiv \frac{3}{\Delta r_i^3}\int dr r^2 f(r),$$

the averaging over θ,

$$\langle f(\theta)\rangle_j \equiv \frac{1}{-\Delta\cos\theta_j}\int d\theta\sin\theta f(\theta),$$

and the averaging over ϕ,

$$\langle f(\phi)\rangle_k \equiv \frac{1}{\Delta\phi_k}\int d\phi f(\phi).$$

(1) $r_{i+1/2} < R$. The field is produced by an inner cell. It follows from (9.1) that

$$-\Phi_{lm}^{i,j,k}(\boldsymbol{R}) = G\Delta m_{i,j,k}\frac{\langle r^l\rangle_i}{R^{l+1}}\langle Q_l^m\rangle_j Q_l^m(\cos\Theta)$$
$$\times\Big(\langle\cos m\phi\rangle_k\cos(m\tilde{\Phi}) + \langle\sin m\phi\rangle_k\sin(m\tilde{\Phi})\Big)$$

and

$$f_{lm}^{i,j,k}(\boldsymbol{R}) = -\frac{\partial\Phi_{lm}^{i,j,k}(R,\Theta,\tilde{\Phi})}{\partial R}$$
$$= -(l+1)G\Delta m_{i,j,k}\frac{\langle r^l\rangle_i}{R^{l+2}}\langle Q_l^m\rangle_j Q_l^m(\cos\Theta)$$
$$\times\Big(\langle\cos m\phi\rangle_k\cos(m\tilde{\Phi}) + \langle\sin m\phi\rangle_k\sin(m\tilde{\Phi})\Big),$$

$$g_{lm}^{i,j,k}(\boldsymbol{R}) = -\frac{1}{R}\frac{\partial\Phi_{lm}^{i,j,k}(R,\Theta,\tilde{\Phi})}{\partial\Theta}$$
$$= G\Delta m_{i,j,k}\frac{\langle r^l\rangle_i}{R^{l+2}}\langle Q_l^m\rangle_j\frac{\partial Q_l^m(\cos\Theta)}{\partial\Theta}$$
$$\times\Big(\langle\cos m\phi\rangle_k\cos(m\tilde{\Phi}) + \langle\sin m\phi\rangle_k\sin(m\tilde{\Phi})\Big),$$

$$h_{lm}^{i,j,k}(\boldsymbol{R}) = -\frac{1}{R\sin\Theta}\frac{\partial\Phi_{lm}^{i,j,k}(R,\Theta,\tilde{\Phi})}{\partial\tilde{\Phi}}$$
$$= mG\Delta m_{i,j,k}\frac{\langle r^l\rangle_i}{R^{l+2}}\langle Q_l^m\rangle_j\frac{Q_l^m(\cos\Theta)}{\sin\Theta}$$
$$\times\Big(-\langle\cos m\phi\rangle_k\sin(m\tilde{\Phi}) + \langle\sin m\phi\rangle_k\cos(m\tilde{\Phi})\Big).$$

After averaging over the volume of the i', j', k' cell, we obtain

$$-\left\langle \Phi_{lm}^{i,j,k} \right\rangle_{i',j',k'} = G\Delta m_{i,j,k} \left\langle r^l \right\rangle_i \left\langle \frac{1}{r^{l+1}} \right\rangle_{i'} \langle Q_l^m \rangle_j \langle Q_l^m \rangle_{j'}$$
$$\times \left(\langle \cos m\phi \rangle_k \langle \cos m\phi \rangle_{k'} + \langle \sin m\phi \rangle_k \langle \sin m\phi \rangle_{k'} \right),$$

$$\left\langle f_{lm}^{i,j,k} \right\rangle_{i',j',k'} = -(l+1)G\Delta m_{i,j,k} \left\langle r^l \right\rangle_i \left\langle \frac{1}{r^{l+2}} \right\rangle_{i'} \langle Q_l^m \rangle_j \langle Q_l^m \rangle_{j'}$$
$$\times \left(\langle \cos m\phi \rangle_k \langle \cos m\phi \rangle_{k'} + \langle \sin m\phi \rangle_k \langle \sin m\phi \rangle_{k'} \right),$$

$$\left\langle g_{lm}^{i,j,k} \right\rangle_{i',j',k'} = G\Delta m_{i,j,k} \left\langle r^l \right\rangle_i \left\langle \frac{1}{r^{l+2}} \right\rangle_{i'} \langle Q_l^m \rangle_j \left\langle \frac{\partial Q_l^m(\cos\theta)}{\partial \theta} \right\rangle_{j'}$$
$$\times \left(\langle \cos m\phi \rangle_k \langle \cos m\phi \rangle_{k'} + \langle \sin m\phi \rangle_k \langle \sin m\phi \rangle_{k'} \right),$$

$$\left\langle h_{lm}^{i,j,k} \right\rangle_{i',j',k'} = mG\Delta m_{i,j,k} \left\langle r^l \right\rangle_i \left\langle \frac{1}{r^{l+2}} \right\rangle_{i'} \langle Q_l^m \rangle_j \left\langle \frac{Q_l^m(\cos\theta)}{\sin\theta} \right\rangle_{j'}$$
$$\times \left(- \langle \cos m\phi \rangle_k \langle \sin m\phi \rangle_{k'} + \langle \sin m\phi \rangle_k \langle \cos m\phi \rangle_{k'} \right).$$

(2) $r_{i-1/2} > R$. The field is produced by an outer cell. In this case, we have

$$-\left\langle \Phi_{lm}^{i,j,k} \right\rangle_{i',j',k'} = G\Delta m_{i,j,k} \left\langle r^l \right\rangle_{i'} \left\langle \frac{1}{r^{l+1}} \right\rangle_i \langle Q_l^m \rangle_j \langle Q_l^m \rangle_{j'}$$
$$\times \left(\langle \cos m\phi \rangle_k \langle \cos m\phi \rangle_{k'} + \langle \sin m\phi \rangle_k \langle \sin m\phi \rangle_{k'} \right),$$

$$\left\langle f_{lm}^{i,j,k} \right\rangle_{i',j',k'} = lG\Delta m_{i,j,k} \left\langle r^{l-1} \right\rangle_{i'} \left\langle \frac{1}{r^{l+1}} \right\rangle_i \langle Q_l^m \rangle_j \langle Q_l^m \rangle_{j'}$$
$$\times \left(\langle \cos m\phi \rangle_k \langle \cos m\phi \rangle_{k'} + \langle \sin m\phi \rangle_k \langle \sin m\phi \rangle_{k'} \right),$$

$$\left\langle g_{lm}^{i,j,k} \right\rangle_{i',j',k'} = G\Delta m_{i,j,k} \left\langle r^{l-1} \right\rangle_{i'} \left\langle \frac{1}{r^{l+1}} \right\rangle_i \langle Q_l^m \rangle_j \left\langle \frac{\partial Q_l^m(\cos\theta)}{\partial \theta} \right\rangle_{j'}$$
$$\times \left(\langle \cos m\phi \rangle_k \langle \cos m\phi \rangle_{k'} + \langle \sin m\phi \rangle_k \langle \sin m\phi \rangle_{k'} \right),$$

$$\left\langle h_{lm}^{i,j,k} \right\rangle_{i',j',k'} = mG\Delta m_{i,j,k} \left\langle r^{l-1} \right\rangle_{i'} \left\langle \frac{1}{r^{l+1}} \right\rangle_i \langle Q_l^m \rangle_j \left\langle \frac{Q_l^m(\cos\theta)}{\sin\theta} \right\rangle_{j'}$$
$$\times \left(- \langle \cos m\phi \rangle_k \langle \sin m\phi \rangle_{k'} + \langle \sin m\phi \rangle_k \langle \cos m\phi \rangle_{k'} \right).$$

(3) $r_{i-1/2} < R < r_{i+1/2}$. Our calculations yield the expressions

$$-\left\langle \Phi_{lm}^{i,j,k} \right\rangle_{i',j',k'} = G\Delta m_{i,j,k} \left(\frac{3}{\Delta r_i^3} \right)^2 I_i^{\Phi l} \langle Q_l^m \rangle_j \langle Q_l^m \rangle_{j'}$$
$$\times \left(\langle \cos m\phi \rangle_k \langle \cos m\phi \rangle_{k'} + \langle \sin m\phi \rangle_k \langle \sin m\phi \rangle_{k'} \right),$$

$$\left\langle f_{lm}^{i,j,k} \right\rangle_{i',j',k'} = G\Delta m_{i,j,k} \left(\frac{3}{\Delta r_i^3} \right)^2 I_i^{1,l} \langle Q_l^m \rangle_j \langle Q_l^m \rangle_{j'}$$
$$\times \left(\langle \cos m\phi \rangle_k \langle \cos m\phi \rangle_{k'} + \langle \sin m\phi \rangle_k \langle \sin m\phi \rangle_{k'} \right),$$

$$\left\langle g_{lm}^{i,j,k} \right\rangle_{i',j',k'} = G\Delta m_{i,j,k} \left(\frac{3}{\Delta r_i^3} \right)^2 I_i^{2,l} \langle Q_l^m \rangle_j \left\langle \frac{\partial Q_l^m(\cos\theta)}{\partial\theta} \right\rangle_{j'}$$
$$\times \left(\langle \cos m\phi \rangle_k \langle \cos m\phi \rangle_{k'} + \langle \sin m\phi \rangle_k \langle \sin m\phi \rangle_{k'} \right),$$

$$\left\langle h_{lm}^{i,j,k} \right\rangle_{i',j',k'} = mG\Delta m_{i,j,k} \left(\frac{3}{\Delta r_i^3} \right)^2 I_i^{2,l} \langle Q_l^m \rangle_j \left\langle \frac{Q_l^m(\cos\theta)}{\sin\theta} \right\rangle_{j'}$$
$$\times \left(- \langle \cos m\phi \rangle_k \langle \sin m\phi \rangle_{k'} + \langle \sin m\phi \rangle_k \langle \cos m\phi \rangle_{k'} \right),$$

where

$$I_i^{\Phi l} = \frac{\Delta r_i^5}{5} \left(\frac{1}{3+l} + \begin{cases} -\dfrac{1}{2-l}, & l \neq 2 \\ \dfrac{1}{5}, & l = 2 \end{cases} \right)$$
$$+ \begin{cases} \dfrac{r_{i+1/2}^5 + r_{i-1/2}^5 - 2\dfrac{r_{i-1/2}^{3+l}}{r_{i+1/2}^{l-2}}}{(3+l)(2-l)}, & l \neq 2, \\ -\dfrac{2}{5} r_{i-1/2}^5 \ln \dfrac{r_{i+1/2}}{r_{i-1/2}}, & l = 2, \end{cases}$$

$$I_i^{1,l} = I_i^{0,l} - 2I_i^{2,l},$$

$$I_i^{0,l} = \frac{r_{i+1/2}^2}{3+l} \left(r_{i+1/2}^2 - r_{i-1/2}^2 \left(\frac{r_{i-1/2}}{r_{i+1/2}} \right)^{1+l} \right) - r_{i-1/2}^2$$
$$\times \begin{cases} \dfrac{1}{2-l} \left(r_{i+1/2}^2 \left(\dfrac{r_{i-1/2}}{r_{i+1/2}} \right)^l - r_{i-1/2}^2 \right), & l \neq 2, \\ r_{i-1/2}^2 \ln \dfrac{r_{i+1/2}}{r_{i-1/2}}, & l = 2, \end{cases}$$

$$I_i^{2,l} = \frac{1}{3+l}\frac{\Delta r_i^4}{4} - \frac{1}{3+l}\begin{cases} \dfrac{r_{i-1/2}^4}{1-l}\left(\left(\dfrac{r_{i-1/2}}{r_{i+1/2}}\right)^{l-1} - 1\right), & l \neq 1 \\ r_{i-1/2}^4 \ln \dfrac{r_{i+1/2}}{r_{i-1/2}}, & l = 1 \end{cases}$$

$$+ \begin{cases} \dfrac{\frac{r_{i+1/2}^4}{2+l}\left(1 - \left(\frac{r_{i-1/2}}{r_{i+1/2}}\right)^{2+l}\right) - \frac{\Delta r_i^4}{4}}{2-l}, & l \neq 2, \\ \dfrac{1}{4}\left[-r_{i-1/2}^4 \ln \dfrac{r_{i+1/2}}{r_{i-1/2}} + \dfrac{\Delta r_i^4}{4}\right], & l = 2. \end{cases}$$

The total acceleration in the i', j', k' zone is

$$\langle \boldsymbol{g} \rangle_{i',j',k'} = \sum_{\substack{i,j,k \\ l,m}} \left\langle \boldsymbol{g}_{lm}^{i,j,k} \right\rangle_{i',j',k'}.$$

The possible sequence for calculating the acceleration $\langle \boldsymbol{g} \rangle_{i',j',k'}$ is as follows:

(1)

$$a_{i,j}^{\binom{1}{2},m} = \sum_k G\Delta m_{i,j,k} \left\langle \begin{matrix} \cos m\phi \\ \sin m\phi \end{matrix} \right\rangle_k, \tag{9.2}$$

(2)

$$b_i^{\alpha,l,m} = \sum_j a_{i,j}^{\alpha,m} \langle Q_l^m \rangle_j, \tag{9.3}$$

(3)

$$c_{i'}^{\alpha,1,l,m} = \sum_{i,i<i'} b_i^{\alpha,l,m} \left\langle r^l \right\rangle_i = c_{i'-1}^{\alpha,1,l,m} + b_{i'-1}^{\alpha,l,m} \left\langle r^l \right\rangle_{i'-1}, \tag{9.4}$$

$$c_{i'}^{\alpha,2,l,m} = \sum_{i,i>i'} b_i^{\alpha,l,m} \left\langle \frac{1}{r^{l+1}} \right\rangle_i = c_{i'+1}^{\alpha,2,l,m} + b_{i'+1}^{\alpha,l,m} \left\langle \frac{1}{r^{l+1}} \right\rangle_{i'+1}, \tag{9.5}$$

$$c_{i'}^{\alpha,\binom{3}{4},l,m} = \left(\frac{3}{\Delta r_{i'}^3}\right)^2 I_{i'}^{\binom{1}{2},l} b_{i'}^{\alpha,l,m}, \tag{9.6}$$

(4)

$$d_{i',j'}^{\alpha,1,m} = \sum_l \Bigg[-(l+1)c_{i'}^{\alpha,1,l,m}\left\langle \frac{1}{r^{l+2}} \right\rangle_{i'} + l c_{i'}^{\alpha,2,l,m}\left\langle r^{l-1} \right\rangle_{i'} + c_{i'}^{\alpha,3,l,m}\Bigg] \langle Q_l^m \rangle_{j'}, \tag{9.7}$$

$$d_{i',j'}^{\alpha,2,m} = \sum_l \left[c_{i'}^{\alpha,1,l,m}\left\langle \frac{1}{r^{l+2}} \right\rangle_{i'} + c_{i'}^{\alpha,2,l,m}\left\langle r^{l-1} \right\rangle_{i'} + c_{i'}^{\alpha,4,l,m}\right] \left\langle \frac{\partial Q_l^m}{\partial \theta} \right\rangle_{j'}, \tag{9.8}$$

$$d_{i',j'}^{\alpha,3,m} = \sum_l \left[c_{i'}^{\alpha,1,l,m}\left\langle \frac{1}{r^{l+2}} \right\rangle_{i'} + c_{i'}^{\alpha,2,l,m}\left\langle r^{l-1} \right\rangle_{i'} + c_{i'}^{\alpha,4,l,m}\right] \left\langle \frac{Q_l^m}{\sin\theta} \right\rangle_{j'}, \tag{9.9}$$

(5)

$$\left\langle \begin{pmatrix} f \\ g \end{pmatrix} \right\rangle_{i',j',k'} = \sum_m \left[d_{i',j'}^{1,\binom{1}{2},m} \langle \cos m\phi \rangle_{k'} + d_{i',j'}^{2,\binom{1}{2},m} \langle \sin m\phi \rangle_{k'} \right], \tag{9.10}$$

$$\langle h \rangle_{i',j',k'} = \sum_m m \left[-d_{i',j'}^{1,3,m} \langle \sin m\phi \rangle_{k'} + d_{i',j'}^{2,3,m} \langle \cos m\phi \rangle_{k'} \right]. \tag{9.11}$$

The quantities $\langle \Phi \rangle_{i',j',k'}$ are calculated in a similar way. This solution is "accurate" in that, under the condition $l_{\max} = j_{\max}$, $m_{\max} = k_{\max}$ ($j_{\max}, k_{\max}$ are the numbers of zones in the θ, ϕ directions, respectively; $l_{\max} \times m_{\max}$ is number of polynomials), the solution is obtained with the same accuracy with witch the density distribution is known. The efficiency of the algorithm, $\propto l_{\max}$ operations per volume point, is fairly high; at least, the integration of hydrodynamic equations using the high-order Godunov-type scheme [108] takes a longer computer time than does the solution of the Poisson equation with $l_{\max} = m_{\max} = 30$ even on a scalar computer. It should be noted that algorithms (9.2)–(9.11) can be easily vectorized. It is very simple and requires less memory than the methods of solving the large system of equations with a sparse matrix that is obtained for a difference approximation of the Laplace operator on the grid. The algorithm also ensures that the boundary condition $\Phi \to 0$ when $r \to \infty$ is satisfied.

In order to calculate $\langle Q_n^m \rangle$, $\left\langle \frac{\partial Q_n^m}{\partial \theta} \right\rangle$, $\left\langle \frac{Q_n^m}{\sin\theta} \right\rangle$, it is convenient to use the expression for P_n^m:

$$P_n^m = \sum_{i=0}^{[n/2]} d_{n,i}^m \begin{cases} \cos(n-2i)\theta, & \text{for even } m, \\ \sin(n-2i)\theta, & \text{for odd } m. \end{cases} \tag{9.12}$$

The coefficients $d_{n,i}^m$ can be directly calculated from the definition of P_n^m:

$$\begin{aligned} P_n^m(\cos\theta) &= (1-\cos^2\theta)^{m/2} \frac{d^m}{(d\cos\theta)^m}\left(\frac{1}{2^n n!}\frac{d^n(\cos^2\theta - 1)^n}{(d\cos\theta)^n}\right) \\ &= \sin^m\theta \sum_{k_0=0}^{[(n-m)/2]} e_{n,k_0}^m \cos^{n-m-2k_0}\theta, \end{aligned}$$

where $e_{n,k}^m = \frac{(-1)^k}{2^n}\frac{(2(n-k))!}{(n-k)!k!(n-m-2k)!}$, and from the equalities

$$\cos^l\theta = \sum_{j=0}^{[l/2]} a_l^j \cos(l-2j)\theta,$$

$$\sin^m\theta = \sum_{k=0}^{[m/2]} b_m^k \begin{cases} \cos(m-2k)\theta, & \text{for even } m, \\ \sin(m-2k)\theta, & \text{for odd } m, \end{cases}$$

where

$$a_l^j = \frac{1}{2^l}\binom{l}{j}\begin{cases} 2, & j \neq l/2, \\ 1, & j = l/2, \end{cases}$$

$$b_m^k = \frac{(-1)^{[m/2]+k}}{2^m}\binom{m}{k}\begin{cases} 2, & k \neq m/2, \\ 1, & k = m/2. \end{cases}$$

The values $\langle Q_n^m \rangle$, $\left\langle \frac{\partial Q_n^m}{\partial \theta} \right\rangle$, $\left\langle \frac{Q_n^m}{\sin\theta} \right\rangle$ must be determined only once, at the beginning of the problem solution. Consequently, the resources that were spent on this calculation have virtually no effect on the solution time of the entire problem.

For a uniform Maclaurin spheroid (a rigidly rotating, incompressible fluid or a rotating star with a very stiff equation of state), only the polynomials P_l^m with $l, m \leq 2$ with $l, m \leq 2$ are involved in the expansion of the potential in terms of spherical harmonics [94]; this will also be the case in the numerical solution provided that the mass distribution is represented fairly accurately on a detailed spatial grid. For this reason, we present the calculation of the potential for a homogeneous sphere ($\rho = 1$) of radius $r_0 = 0.5$ centered at point $(r, \theta, \phi) = (r_0, \pi/2, 0)$ as a more substantial test for the numerical solution of the Poisson equation. The analytical solution (we set $G = 1$) is

$$\Phi = -M \begin{cases} \left(3 - r^2/r_0^2\right)/2r_0, & \eta \leq r_0, \\ 1/\eta, & \eta > r_0, \end{cases}$$

where $M = (4\pi/3)r_0^3$ and $\eta = \sqrt{r^2 + r_0^2 - 2rr_0 \sin\theta\cos\phi}$. The calculation is performed on a $20 \times 40 \times 40$ grid in the region $0 \leq r \leq 1$, $0 \leq \theta \leq \pi$, $0 \leq \phi \leq 2\pi$ with various numbers of polynomials P_l^m ($l \leq l_{\max}$, $m \leq m_{\max}$). Figure 9.1 shows the functions $\Phi(x, y = 0, z = 0)$ and $\Phi(x = 0, y = 0, z)$, where $x = r\sin\theta\cos\phi$, $y = r \sin\theta\sin\phi$, $z = r\cos\theta$, for the calculations with $l_{\max} = m_{\max} = 2, 5, 20$. We see that satisfactory results are obtained even for $l_{\max} = m_{\max} = 5$. The gravitational acceleration $\mathbf{g} = -\text{grad}\,\Phi$ is commonly determined by differentiating the potential specified on a grid. Since our method involves no numerical differentiation, the error in the acceleration is the same as that in the potential (see Fig. 9.1). For checking purposes, Fig. 9.2 also shows the contour lines of Φ in the $z = 0$ and $y = 0$ planes; they must be the circumferences centered at points $(x, y) = (0.5, 0)$ and $(x, z) = (0.5, 0)$, respectively.

It should be noted that the largest relative error in the potential turns out to be near the center of the sphere (Fig. 9.1). Since the equation and the numerical solution are linear with respect to ρ, the relative error will be smaller if there is a second identical mass centered, for example, at point $(x, y, z) = (-0.5, 0, 0)$. In particular, our algorithm proves to be suitable for determining the gravitational field when calculating the evolution of a binary system of neutron stars.

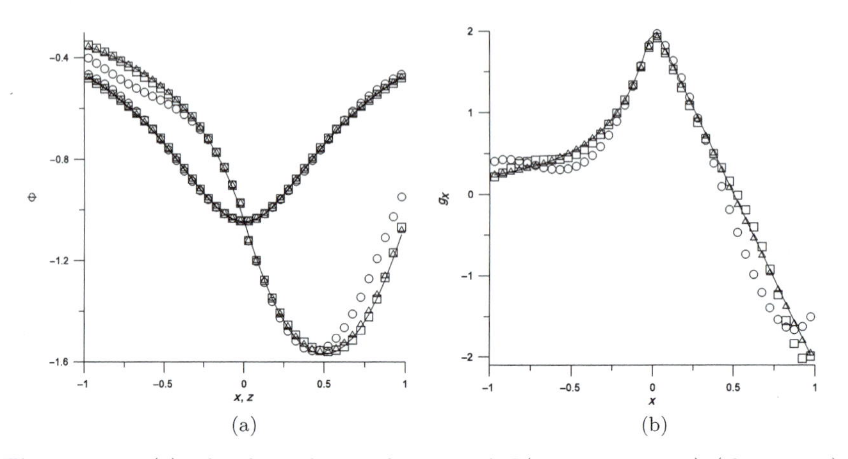

Figure 9.1 (a) The dependence of potential $\Phi(x, y = 0, z = 0)$ (thin curve) and $\Phi(x = 0, y = 0, z)$ (thick curve). (b) The dependence of gravitational acceleration $g_x(x, y = 0, z = 0)$ for a homogeneous sphere of radius $r_0 = 0.5$ centered at point $(x, y, z) = (0.5, 0, 0)$. The curves represent analytical solutions; the circles, squares, and triangles are numerical solutions with $l_{\max}, m_{\max} = 2, 5, 20$, respectively [10].

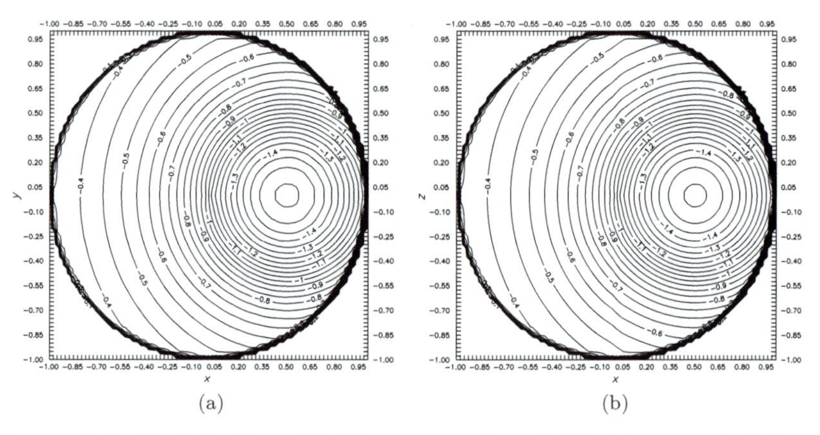

Figure 9.2 Contour lines (a) $\Phi(x, y, z = 0)$ and (b) $\Phi(x, y = 0, z)$ for a homogeneous sphere of radius $r_0 = 0.5$ centered at point $(x, y, z) = (0.5, 0, 0)$, $l_{\max}, m_{\max} = 20$. The calculation is performed in the region $r < 1$; $\Phi = 0$ at $r > 1$ [10].

The method described above was used to analyze the stability of rotating polytropes and neutron stars in 2D and 3D cases [6,24] to study fragmentation and to calculate the gravitational radiation during fragmentation of a rapidly rotating neutron star [9].

Part 3

Applications

Chapter 10

The Problem of the Regime of Burning in Degenerate Carbon–Oxygen Cores in Type I Supernovae

An attempt is made to solve the problem of the regime of burning in degenerate carbon–oxygen cores (CO cores) of stars, which are assumed to be Type Ia presupernovae, in favor of deflagration burning. For this purpose, we demonstrate numerically for *Chapman–Jouguet detonation* waves the galloping instability that changes into decay. The structure of these waves is described by the Zel'dovich–Neumann model involving the kinetics of nuclear reactions in the burning zone. Along with the numerical hydrodynamic model, which includes the kinetics of nuclear reactions, we suggest a qualitative model of *galloping instability*, which takes nonlinear perturbations into account. It is shown that the decay takes a very short time ($\lesssim$ 10 ms) and is nearly independent of the initial disposition of detonation waves along the radius of the CO core. The kinetic equations include 12 nuclides, from ^{12}C to ^{28}Si, with fairly detailed net of nuclear reactions (39 reactions). The initial composition of the CO core is an equimolar mixture of ^{12}C and ^{16}O (with equal mass fractions), while the mass of the CO core is close to the Chandrasekhar limit. In addition to the mixture of Boltzmann nuclide gases, the equation of state takes into account the *relativistic electron–positron gas* of arbitrary degree of degeneracy and the equilibrium radiation.

10.1 Introduction

It is generally agreed that explosions of Type Ia supernovae (SNIa) are associated with thermonuclear burning in degenerate CO cores of supernovae. These explosions are accompanied by the formation of heavy elements, and the central problem of the theory is to determine the fraction of these elements in the total mass of the burnt matter. In order to answer this question, it is necessary to start with the understanding of the regime in which the burning of matter occurs in Type Ia supernovae. This may be either supersonic *detonation* or *deflagration*. To the latter we also refer spontaneous burning, which may occur both in the subsonic and in supersonic modes.

The detonation regime of propagation of the burning front in CO cores was examined in Ref. [37]. However, the basic problems connected with the overproduction of iron in SNIa explosions were raised in that study.

In a series of papers by Ivanova *et al.* [180,182,184,185] investigated a model of thermonuclear burning in CO cores, which takes into account the finite rate of carbon burning. It was found that the burning of matter occurs, in general, in the *subsonic deflagration regime* and that the propagation of the burning front is accompanied by pulsations of the CO core. Since heat conduction and/or convection is not included in this model, such a burning regime could be attributed to spontaneous regimes.

The authors [296] considered the problem of propagation of the deflagration wave in the *degenerate CO* core. The wave propagation was caused by the finite heat conduction of matter. According to this work, the velocity of the burning wave front is essentially subsonic and is insufficient for explaining some observational data. Attempts therefore were made in a few studies to ascertain the possible mechanisms for acceleration of the burning front. These mechanisms are related to the development of various instabilities: *diffusion–thermal instability* [89], *Rayleigh–Taylor geometrical instabilities* [191,243,244], and *Landau instability* [77]. However, the question of the efficiency of such acceleration remains open.

The detonation and deflagration regimes are not mutually exclusive: Mixed regimes are also possible. Different regimes of burning can be realized at different stages of SNIa explosion. In such cases, either the deflagration wave initiates detonation or the detonation wave decays and changes into deflagration. It is evident that the transition of one burning regime into another can be considered only if we take into account the finite rate of nuclear reactions. For example, Ivanova *et al.* in the above-mentioned paper [185] demonstrated the transition of deflagration into detonation.

The next work of this kind, as applied to the SNIa theory, was [76], in which the transformation of the spontaneous burning wave into the detonation wave was examined. It was shown that such a process is quite probable in degenerate CO cores. It ends in the formation of the detonation wave at the presupernova center, in the region of very small radius ($\sim 10^5$ cm). The mass embraced by the detonation wave at the time of its formation is $\sim 6.5 \cdot 10^{-9}\, M_{\odot}$. Further evolution of this detonation wave is unclear: Does it travel in the steady-state or galloping regime or does it even decay into the deflagration wave plus the shock wave that runs away from the deflagration wave?

In this study, we will attempt to answer this question. As we noted above, the answer can be found, if we take into account the finite rate of nuclear reactions. We disregard the heat conduction and/or convection and interpret the contact discontinuity, which arises in the form of the burning zone in the case of the detonation decay, as a deflagration wave front.

10.2 Qualitative analysis

The detonation wave is considered in the following framework of the *Zeldovich–Neumann plane steady-state model* [199]. This model suggests that the burning mixture is first compressed in the shock wave without changing its composition and only then nuclear reactions begin behind the shock wave front. In the case of a plane stationary structure of the detonation wave with the fixed energy release Q per unit mass, variations in thermodynamic parameters of the matter are usually represented by the (V, p) diagram (Fig. 10.1), where V is the specific volume and p is the pressure.

Let point O on the diagram represent the matter located ahead of the front of the detonation wave. The state of this matter after passing through the shock wave front is marked by point A, which lies on the shock adiabatic curve OA. During the burning process, the imaging point then descends along the Rayleigh–Michelson straight

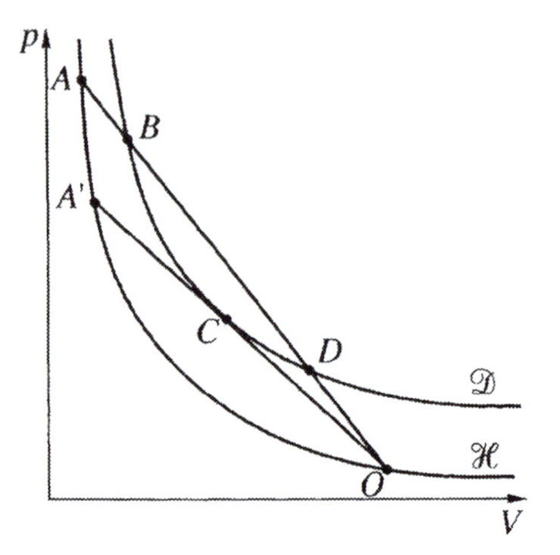

Figure 10.1 Diagram of thermodynamic states of the gas: $\mathcal{H}$ is the shock adiabatic curve, $\mathfrak{D}$ is the detonation adiabatic curve, and C is the Chapman–Jouguet point. The combination of straight lines OA–AB represents the regime of overcompressed detonation. The greater the difference in the slopes of lines OA and OC, the higher the degree of overcompression. The combination of straight lines OA'–$A'C$ describes the regime of spontaneous Chapman–Jouguet detonation. The reaction zone is matched by the segments of straight lines AB and $A'C$, respectively [172].

line until point B in the detonation adiabatic curve, where the burning stops, is reached. The slope of the Rayleigh–Michelson straight tine is the squared density (with inverse sign) of the mass flow that passes through the wave front. Figure 10.1 shows the existence of the minimum rate of the steady-state detonation wave. This rate corresponds to the tangent to the detonation adiabatic curve which comes out of point O. Such regime of the wave propagation is called the Chapman–Jouguet spontaneous detonation. In general, we show in Fig. 10.1 the regime of overcompressed detonation.

This steady-state model describes the traveling detonation wave independently of the concrete kinetics of reactions that occur in the burning zone, which corresponds to segment AB on the diagram. The kinetics determines only the internal structure of the reaction zone, i.e., the spatial distribution of concentrations and other parameters behind the shock wave front. It is important, however, that reactions at each burning stage occur with energy release. Otherwise, a more complicated structure may form, namely, the intrinsic detonation, whose burning zone corresponds to segment AD on the diagram.

The authors [173] studied the plane steady-state internal structure of the detonation wave in the CO core using a realistic equation of state and a detailed net of nuclear reactions. It was shown that the transition from the initial chemical composition to the state of statistical nuclear equilibrium immediately behind the front of the bow shock (pure carbon) occurs in three steps: (i) carbon burning, (ii) transition to the statistical quasi-equilibrium state, and (iii) transition to the state of perfect statistical equilibrium. The thickness of the burning zone depends very strongly on density, which makes impossible, in particular, the formation of the steady-state detonation wave at the edge of the CO core at low densities ($\lesssim 10^7$ g cm^{-3}): The burning zone becomes thicker than the distance to the edge. In this study, we have also investigated the structures of overcompressed detonation waves.

The authors [127,261] studied the instability of detonation waves in the framework of the linear perturbation theory in the Zeldovich–Neumann model with one reaction. These studies enable one to make the following, possibly general, statements concerning stability:

- Detonation waves are more stable at a high degree of overcompression (Fig. 10.1).
- Instability with respect to spatial perturbations sets in earlier than instability with respect to 1D perturbations.

The problems of *stability of the detonation wave* under presupernova conditions are considered in Refs. [191,193]. It was shown that the detonation wave, if it forms, is unstable in almost all cases and either decays or spreads in the multifront mode (see the following). As to the statement "unstable in almost all cases", we should make the following remark. This conclusion, made in the second of the above papers, is based on the fact that the detonation wave travels virtually through the entire mass of the presupernova in the Chapman–Jouguet regime, which is most sensitive to various instabilities.

If, for some reason, the detonation wave propagates in the overcompressed regime, these instabilities can be stabilized [191]. One of the stabilizing factors is a multistep burning. The authors [193] examined, for example, the kinetics involving a single reaction, $C^{12}+C^{12} \to Ne^{20}+\alpha$, with the energy release $Q_1 = 0.18\cdot 10^8$ erg g^{-1}. In reality, as a result of other nuclear transformations, the energy release Q_2 may

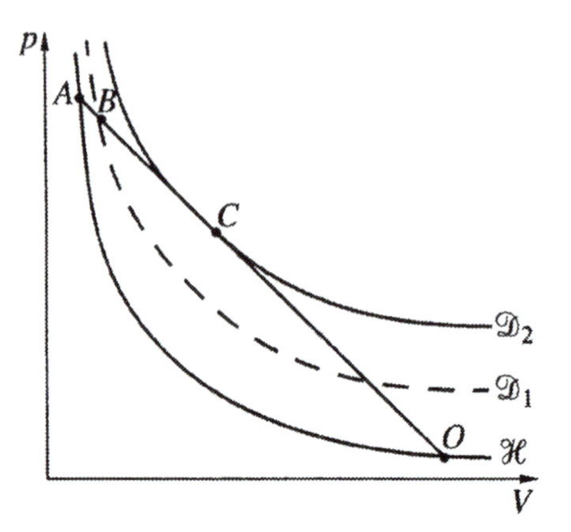

Figure 10.2 Diagram of thermodynamic states of the gas: $\mathcal{H}$ is the shock adiabatic curve and $\mathcal{D}_1$ and $\mathcal{D}_2$ are the detonation adiabatic curves constructed according to the energy release in the reaction $C^{12} + C^{12} \to Ne^{20} + \alpha$ (with the caloricity Q_1) and to the total energy release (with the caloricity Q_2, where $Q_1 < Q_2$). Rayleigh–Michelson straight line AO corresponds to Chapman–Jouguet regime for the adiabatic curve $\mathcal{D}_2$ and to the overcompressed detonation for the adiabatic curve $\mathcal{D}_1$ [172].

be appreciably higher. The wave calculated for this higher-energy release and traveling at the Chapman–Jouguet velocity will be overcompressed for the (shock wave) and ($C^{12}+C^{12} \to Ne^{20}+\alpha$ reaction zone) structure, which was interpreted in the cited paper as a spontaneous detonation wave. This assertion is illustrated in Fig. 10.2. It is easy to understand that, if the development of instability for the wave with high-energy release results in the cessation of burning at an earlier stage, further development of instability may stop.

The spontaneously propagating Chapman–Jouguet wave is accompanied, as a rule, by the adjacent rarefaction wave, which is by no means a simple self-similar wave due to its non-uniformity, non-stationarity, and sphericity in the burnt matter. Since, in this case, the gas velocity behind the wave, as measured with respect to its front, is equal to the speed of sound, there is no interaction between these waves. However, if the detonation wave spreads in the *non-stationary galloping regime* (by definition, in the case of the galloping regime, the burning zone oscillates together with the shock wave in such a way that the latter spreads by jerks with variable velocity), the interaction between the detonation wave (which is understood to be the (shock wave) + (reaction zone) system) and the rarefaction wave may become essential. We can assert that the propagation of the detonation wave in the CO core (if this is the case) occurs precisely in the galloping mode. In our study, we investigated the interaction indicated above in the degenerate matter of the CO core.

The pressure in the degenerate matter depends largely on density. Both the heat conduction (at constant volume or pressure) are small, and even slight variations in pressure, density, and element concentrations lead to considerable temperature variations. As usual, the nuclear reaction rates are very sensitive to temperature. This leads to the situation, where the main energy release occurs at the end of the burning zone (at least, for small wave overcompression, ~1), and, therefore, along with the shock wave front, a narrow burning front also exists. The structure including the shock front and the burning front, separated by the induction zone, is called the two-front detonation, which cannot exist in the steady-state condition.

To justify the last assertion, let us consider the (u, p) diagram, which illustrates the basic possibility for the decay of the two-front detonation. Here, u is the hydrodynamic velocity (see Fig. 10.3 and the work [167]).

Let a two-front detonation wave propagate along a "background", which is represented by point O. The gas in this wave is compressed by the shock wave to the state marked by point A and then, after passing through the burning front, goes over into the state denoted by point B. Points O, A, and B lie in the single Rayleigh–Michelson

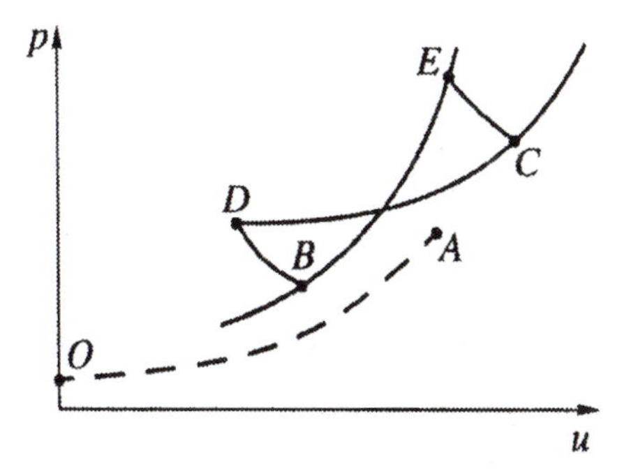

Figure 10.3 The (u, p) diagram of decay of the two-front detonation wave. OA and BE are, respectively, the shock and detonation adiabatic curves for state O; C is the Chapman–Jouguet point for state A, which lies on the corresponding detonation adiabatic curve; BD and CE are the shock adiabatic curves of points B and C, respectively [172].

straight line that is not shown here for simplicity. The shock wave (in state A) can be followed by detonation waves. From point A on the (u, p) diagram, one can arrive at point C, after passing through the Chapman–Jouguet wave. Point C lies in the detonation adiabatic curve constructed for point A. The detonation wave also has a two-front structure, which is not considered here. The Chapman–Jouguet wave is accompanied by the adjacent rarefaction wave, the gas condition in which is represented by line CD. In this case, the shock wave, in which the gas undergoes the sharp transition $B \to D$, spreads through the burnt gas in the opposite direction. The process described here is represented by the (x, u) diagram in Fig. 10.4.

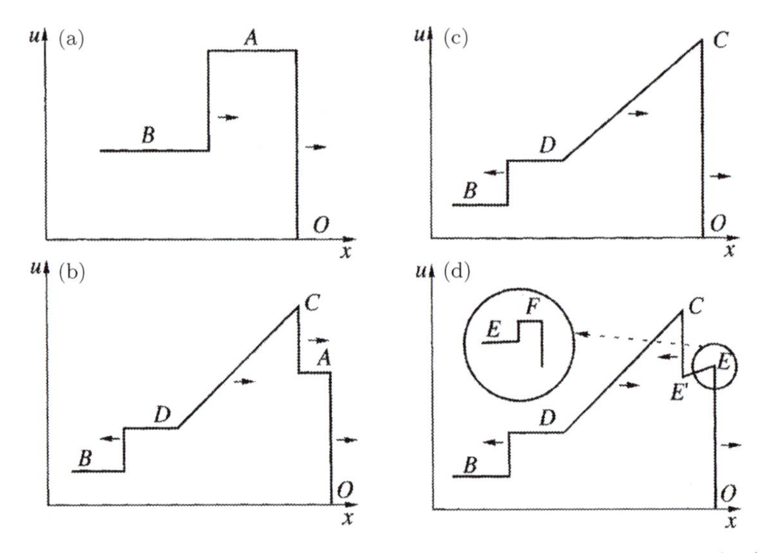

Figure 10.4 The scheme of decay of the two-front detonation wave on the (x, u) plane: (a) the initial steady-state structure of the detonation wave: the sharp change $O \to A$ corresponds to the bow shock wave; the sharp change $A \to B$ corresponds to the burning front; (b) the structure of the detonation wave after the decay of the burning front: $O \to A$ is the bow shock wave; $A \to C$ is the Chapman–Jouguet wave; $C \to D$ is the rarefaction wave following Chapman–Jouguet wave; $B \to D$ is the shock wave traveling through the burnt matter to the left; (c) the flow structure at the time of the onset of interaction between the bow shock and Chapman–Jouguet wave that overtook the bow shock; (d) the flow structure arising after the above interaction: $O \to E$ is the overcompressed detonation wave, whose internal structure contains the burning front $F \to E$ (shown on enlarged scale); $C \to E$ is the shock wave which results from the decay of the abrupt change $O \to C$ and which travels to the left [172].

Figure 10.4(a) shows the initial structure of the two-front detonation wave, while the structure after the decay of the burning front is shown in Fig. 10.4(b). Since the flow behind the bow shock ($O \to A$) is subsonic, the Chapman–Jouguet ($A \to C$) wave overtakes the bow shock after a time. The merging of these waves causes a overcompressed detonation wave to form. When passing through the detonation wave, the condition of the gas changes abruptly, $O \to C$, at the instant of time corresponding to the onset of interaction (Fig. 10.4(c)). Further, as the overcompressed detonation wave spreads through the gas, a complicated flow pattern forms behind it (Fig. 10.4(d)).

The energy release in the reaction zone of the overcompressed detonation wave occurs relatively uniformly; as a result, at high overcompression, this wave is progressively attenuated due to the interaction with the rarefaction wave that follows the detonation wave. When the detonation wave is sufficiently slowed down, and the energy release is confined to the narrow layer, a new decay cycle occurs. The cyclic process of this kind was not revealed, however, in numerical calculations.

Qualitative considerations on the character of propagation of the detonation wave in the degenerate matter can be verified only by numerical modeling of gas-dynamical processes with allowance for the equation of state and the kinetics of nuclear reactions.

10.3 Statement of the problem and computing algorithm

We assume that the detonation wave, formed at a certain time at the presupernova center, travels then in the spherically symmetric way toward the periphery. The wave formation process is not considered here (it may develop, for example, according to the mechanism suggested in Ref. [76]): The spontaneous detonation wave is assumed to arise and spread with the speed $U = U_{\mathrm{CJ}}$. We also assume that this wave has the plane steady-state structure that we discussed above.

The steady-state condition means that in the coordinate frame related to the wave front moving at the speed U, the structure of the reaction zone does not vary with time. This steady-state structure is described by the set of ordinary differential equations:

$$(U - u)\frac{dY_i}{dx} = W_i, \tag{10.1}$$

where x is the 1D coordinate of the plane detonation wave, which is subject to the Hugoniot conditions in the entire reaction zone:

$$\begin{aligned}-U(\rho-\rho_0)+\rho u&=0,\\ -U\rho u+(\rho u^2+p)-p_0&=0,\\ -U\left(\rho\frac{u^2}{2}+\rho E-\rho_0E_0\right)+u\left(\rho\frac{u^2}{2}+\rho E+p\right)&=0.\end{aligned}\tag{10.2}$$

In Eqs. (10.2), x is the distance from the front of the bow shock, ρ is the density, p is the pressure, u is the speed of matter, and $E=\epsilon+\sum Q_iY_i$. Here, ϵ is the specific internal energy of the matter, $\sum Q_iY_i$ is the specific heat energy release, $Y_i = N_i/(\rho N_A)$ is the molar fraction (abundance) of the element with atomic number i, N is the number of nuclide particles of species i per unit volume, Q_i is the binding energy of the nuclide of species i multiplied by the Avogadro number N_A, and W is the rate of formation (exhaustion) of the ith element in all reactions. The subscript 0 indicates the values of these quantities ahead of the wave front. We disregard here the curvature of the front and the gravitational force, because, in general, the width of the burning front is much smaller than the core radius, while the energy release is sufficiently high.

A complicated rarefaction wave, in which the speed of matter reduces to zero at the center of the star, is adjacent to the reaction zone. The subsequent spherically symmetric propagation of this detonation wave is described by the gas-dynamic equations with burning kinetics:

$$\begin{aligned}\frac{\partial\rho}{\partial t}+\frac{1}{r^2}\frac{\partial(r^2\rho u)}{\partial r}&=0,\\ \frac{\partial(\rho u)}{\partial t}+\frac{1}{r^2}\frac{\partial}{\partial r}\left(r^2\left(\rho u^2+p\right)\right)&=\frac{2p}{r}-\frac{GM}{r^2}\rho,\\ \frac{\partial}{\partial t}\rho\left(\frac{u^2}{2}+E\right)+\frac{1}{r^2}\frac{\partial}{\partial r}\left(r^2\rho u\left(\frac{u^2}{2}+E+\frac{p}{\rho}\right)\right)&=-\frac{GM}{r^2}\rho u,\\ \frac{\partial(\rho Y_i)}{\partial t}+\frac{1}{r^2}\frac{\partial(r^2\rho uY_i)}{\partial r}&=\rho W_i,\end{aligned}\tag{10.3}$$

where M is the mass of the matter confined within the sphere of radius r, all other quantities being defined above.

Equations (10.3) are therefore considered in the entire region bounded by the bow shock wave, i.e., for $0 \leq r \leq R(t)$, where $R(t)$ determines the position of the shock wave front.

The study of thermonuclear burning in the CO core calls for the "universal" equation of state, which can be used for a wide range of temperatures and densities. In addition to nuclides, which obey the equation of state of ideal gas, we take into account the contribution of electrons and positrons, as well as the equilibrium radiation. The expressions for pressure and specific internal energy are

$$p = \rho N_A k_{\mathrm{B}} T \sum_i Y_i + \frac{aT^4}{3} + P_{\pm},$$

$$E = \frac{3}{2} N_A k_{\mathrm{B}} T \sum_i Y_i + \frac{aT^4}{\rho} + \epsilon_{\pm} + \sum_i W_i Y_i,$$

where k_{B} is the Boltzmann constant, Y_i is the abundance of the element with atomic number i, a is the constant of radiation energy density, $p_{\pm} = p_+ + p_-$ is the pressure, and $\epsilon_{\pm} = \epsilon_+ + \epsilon_-$ is the specific internal energy of the electron–positron gas.

For the real computation, we subdivided the region of thermodynamic variables, 10^7 g cm$^{-3} \leq \rho/\mu_e \leq 10^{10}$ g cm^{-3}, 10^7 K $\leq T \leq 10^{11}$ K, into the four subregions (μ_e is the mean molecular weight per electron, equal to 2 in our case), where different asymptotic expansions are used for thermodynamic functions of the electron–positron gas (see, e.g., Ref. [75]).

The boundaries of conjugation between the above four subregions were selected in such a way that the continuity of first derivatives is satisfied.

In most variants, the following net of nuclear reactions was considered:

$$^{12}\mathrm{C} + {}^{12}\mathrm{C} \to {}^{20}\mathrm{Ne} + \alpha,$$

$$^{12}\mathrm{C} + \alpha \to {}^{16}\mathrm{O},$$

$$^{16}\mathrm{O} + \alpha \to {}^{20}\mathrm{Ne}.$$

The inverse reactions were disregarded. This net of equations yielded, in passing from ^{12}C to ^{20}Ne, a caloricity of $0.18 \cdot 10^{18}$ erg g^{-1} = 0.19 MeV per nucleon.

The extended net of nuclear reactions containing 39 reactions for nuclides ^{12}C, ^{13}C, ^{16}O, ^{20}Ne, ^{23}Na, ^{24}Mg, ^{26}Al, ^{28}Si, ^{30}Si, α particles, and protons with a caloricity of $0.32 10^{18} \mathrm{erg\,g^{-1}} = 0.32$ MeV per nucleon was considered in Ref. [172].

Since the computation facilities did not permit us to follow the evolution of the detonation wave from its generation to decay, or to its break-out at the boundary of the CO core, we assumed that the detonation wave spreads in the Chapman–Jouguet regime up to the radius R_0, which is the parameter of the problem. Beginning from the time the front of the detonation wave reaches the radius R_0, we performed the calculation of the complete set of Eqs. (10.3), with the kinetics of nuclear reactions indicated above.

The initial data for this set of equations were specified as follows. The computation region, with the left boundary placed at the core center and the right edge positioned at the shock front R_p, was divided into several thousands of computation cells. The computation began with resolving the Hugoniot conditions at the shock wave (10.2) with the known speed U and quantities ρ_0, p_0, and E_0. This enabled us to determine the density and temperature values behind the shock wave and, correspondingly, the nuclear reaction rates. In each cell, we solved the set of ordinary differential equations (10.1) from R_0, toward the center. In each subsequent computation cell, the density and temperature values were corrected as a function of the energy release, according to conditions (10.2). The set of equations (10.1) was solved by the implicit linear multistep method, the choice of which was determined by the high degree of stiffness. Clearly, this choice of the initial conditions has no physical meaning of its own because the Hugoniot conditions (10.2), derived for plane geometry, are extended artificially to the spherically symmetric case, with the left edge of the computation region at the core center. However, it is clear that the basic change of the quantities Y_i occurs in the small vicinity of point R_0, the right edge of the computation region, where sphericity is a small factor. In the remaining part of the computation region, the quantities Y_i are constant and therefore the remaining quantities p, u, and T are also constant.

As a result of solving the ordinary differential equations (10.1) with the Hugoniot conditions (10.2), we obtain the distributions of the functions $Y_i(x)$, $p(x)$, $u(x)$, and $T(x)$ in the region from a point behind the shock wave front, where $x = 0$, to the center of the CO core. These distributions are the initial conditions for the non-stationary set of equations (10.3).

To solve the set of equations (10.3) in the region $0 < r < R(t)$, we introduced the moving computation grid $\{r_n(t), n = 0, \ldots, N\}$, which enabled us to resolve the burning zone with the width $\delta \ll R$. The extreme left node is located at the core center: $r_0 = 0$; the node at the extreme right is at the front of the bow shock: $r_N = R(t)$ (at the initial time, $R(O) = R_0$).

To numerically integrate equations (10.3), we used the Godunov-type difference scheme. All hydrodynamic quantities (density, velocity, etc.) were assigned to the grid intervals. The motion of the grid nodes with velocities dr_n/dt leads to the correction of the flows F through surfaces $r = r(t)$: $F_n \to F_n - f_n dr_n/dt$, where f_n is the corresponding density at the node n. Here, f denotes the mass density, momentum density, energy density, and nucleus number density $\propto \rho Y_i$.

To determine the velocity of the bow shock, we solved the problem of decay of the discontinuity between the background state and the state behind the shock wave (the latter state was taken to be the state in the extreme right cell).

The following procedure was used to construct the computation grid and calculate the velocities of nodes. We introduced the variable $q = \sum_i Q_i(Y_i^0 - Y_i)$, where Y_i^0 are the background values of nuclide abundances and Y_i are the local abundance values. The quantity q has an obvious meaning of the total specific heat energy that is released in changing from the state with abundances Y_i^0 into the state with abundances Y_i. We fix the spatial point N_*, where the specific concentration $12Y_{\mathrm{C}}$ of burnt carbon is 0.2 (the initial value is $12Y_{\mathrm{CO}} = 0.5$) and the corresponding total energy release is q_*. The value of q_* is of the order of the total caloricity of nuclear reactions (but less than this value). The separation between the node, $n = N_* < N$ located at $q = q_*$, and the bow shock, where $q = 0$ and $n = N$, is taken to be the width δ of the reaction zone. The latter is divided into $N - N_*$ difference intervals uniformly with respect to the variable q. Such an approach makes it possible to shrink the difference grid in sites of strong energy release.

The velocities of nodes were chosen so as to ensure the conservation of q along the trajectories of nodes, i.e.,

$$\frac{\partial q}{\partial t} + \frac{dr}{dt}\frac{\partial q}{\partial t} = 0.$$

This condition gives, with allowance for the obvious relation

$$\frac{\partial q}{\partial t} = -u\frac{\partial q}{\partial t} + \sum_i Q_i W_i,$$

which follows from the kinetic equations (10.3), the equation

$$\frac{\partial r}{\partial t} = u - \frac{\partial r}{\partial q}\sum_i Q_i W_i. \tag{10.4}$$

The relation (10.4) was used only in the reaction zone, i.e., in the region $0 < q < q_*$, $R - \delta < r < R$. In the remaining region, $0 < r < R - \delta$, where the energy release is low, we constructed the uniform grid, using the geometric progression, in such a way that the ratio of lengths of neighboring intervals (at the fixed time) was constant: $\dfrac{\Delta r_{n+1/2}}{\Delta r_{n-1/2}} = \lambda(t)$ and $\Delta r_{N_*-1/2} = \Delta r_{N_*+1/2}$.

10.4 Results of calculations

We recall that the plane steady-state structure of the *Chapman–Jouguet detonation wave* is calculated at the initial time, using (10.1) and (10.2). Only at the next instants of time we take into account of the spherical, non-uniform, and non-stationary character of the wave, according to (10.3), and the structure of the detonation wave ceases to be steady-state and plane.

A series of computations were carried out within the framework of the problem described above. For four such computations, the following data are presented in Table 10.1: the computation number (No), the parameters of the starting point of the detonation wave (the density ρ_0, the pressure p_0, and the speed of sound c_0 ahead of the wave front), and the central density ρ_c (the temperature $T_c = 3 \cdot 10^8$ K). These parameters are used to calculate the equilibrium configuration of the CO core. Moreover, for each computation, we present in the

Table 10.1 Parameters of tasks.

No.	ρ_0 10^9 g cm^{-3}	p_0 10^{27} dyn cm^{-2}	c_0 10^9 cm s^{-1}	ρ_c 10^9 g cm^{-3}	U_{CJ} 10^9 cm s^{-1}	ρ_{CJ}/ρ_0 10^9 g cm^{-3}	f
1	2	1.2	0.914	2	1.27	1.26	0.86
2	1	0.46	0.813	2	1.18	1.28	0.82
3	0.005	0.00032	0.312	2	0.75	1.54	0.92
4	5	4.2	1.065	5	1.42	1.47	0.86

table the initial velocity of the detonation wave U_{CJ} and the initial compression at the detonation wave front ρ_{CJ}/ρ_0, as well as, the final degree of deceleration of the shock wave, $f = U/U_{\text{CJ}}$. It should be stressed that the quantity U_{CJ}, which appears in the definition of f, is assumed to be constant and equal to its initial values presented in the table. All values in the table are given in the CGS system of units.

Using the data presented in the table, we were able to reproduce, with allowance for *Hugoniot conditions* (10.2), all other quantities behind the *Chapman–Jouguet detonation wave* that appear in the initial conditions, in particular, c_{CJ}, p_{CJ}, and the velocity $u_{\text{CJ}} = c_{\text{CJ}} + U_{\text{CJ}}$ for this wave (the determination of the specific internal energy and the corresponding temperature calls for the use of the equations of the state indicated above). Note that the Chapman–Jouguet wave was strong only in variants 3, i.e., we can ignore the pressure ahead of the wave front and obtain, according to Ref. [200], $U_{\text{CJ}} = \sqrt{2 \cdot 1.78q} \cdot 0.8 \cdot 10^9$ cm s^{-1}. Here we used the adiabatic exponent $\gamma = 5/3$, whereas, according to the equation of state, the adiabatic exponent is slightly lower.

We considered the degenerate CO core of the presupernova with mass close to the Chandrasekhar limit $M = 1.4\,M_\odot$. The chemical composition (expressed in specific number density) was 50% of ^{12}C and 50% of ^{16}O, in terms of specific concentrations. The temperature was assumed to be constant in the radial direction and equal to $T_c = 3 \cdot 10^8$ K in all variants. Computations 1–3 were carried out for the central density $\rho_c = 2 \cdot 10^9$ g cm^{-3} with the standard set of reactions, differing only by the starting point of the detonation wave. The density, pressure, and speed of sound ahead of the detonation wave are specific for each variant, while the temperature T_c is the same in all variants. Also, we performed the additional computation 4 that differs from the previous computations by the higher value of the central density of the equilibrium configuration, $\rho_c = 2 \cdot 10^9$ g cm^{-3}, which is still justified for equilibrium configurations of CO cores in binary systems [180,182,184].

In all variants we obtained the development of the galloping instability of the detonation wave, which proceeds with increasing amplitude of oscillations (in terms of the degree of deceleration of the shock wave and the width of the reaction zone) and ends in the decay of detonation. In this case, the bow shock breaks away from the reaction zone, which in turn degenerates into the "blurred" contact discontinuity. Figure 10.5 shows the degree of deceleration f of the detonation wave and the width δ of the reaction zone for variant 3 as a function of time. After the decay of detonation, the quantity δ grows infinitely, while the parameter $f \longrightarrow c_0/U_{\rm CJ}$, where c_0 is

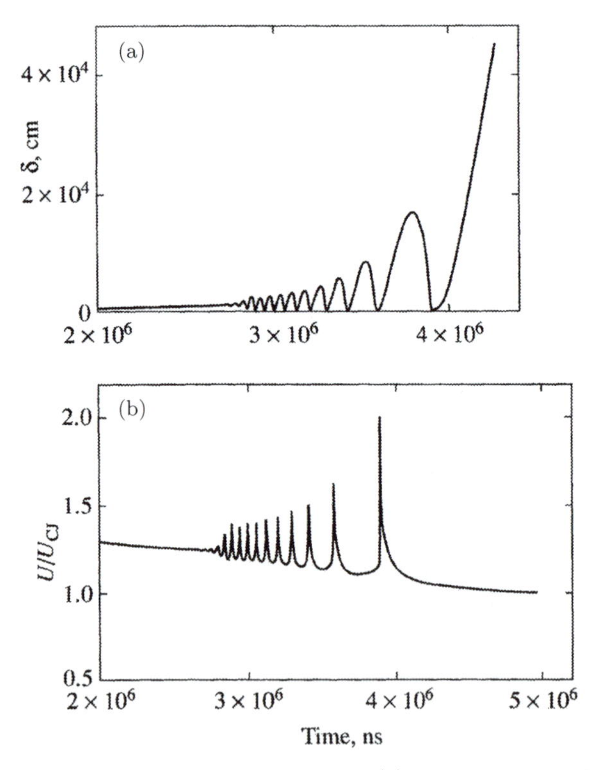

Figure 10.5 The width δ of the reaction zone (a) and the degree of deceleration $f = U/U_{\rm CJ}$ of the shock wave (b) for variant 3 as a function of time [172].

the speed of sound ahead of the wave. Note that the initial value of the parameter f in all variants is equal to unity (Chapman–Jouguet detonation wave).

In other variants, the behavior of the functions $f(t)$ and $\delta(t)$ is qualitatively similar, differing from the behavior described above by the numerical values of the period and amplitude of oscillations.

We also note a strong dependence of the time of cessation of oscillations of the width of the reaction zone on the computation parameters (chiefly, on the starting point of the detonation wave). For the detonation wave starting from the center of the star, this time is ~10–50 ns. As the starting point moves away from the center, this time increases to ~200 ns (for $\rho_0 = 10^9$ g cm^{-3} and $R_0 = 4.8 \cdot 10^7$ cm) and reaches $5 \cdot 10^6$ ns (for $\rho_0 = 5 \cdot 10^6$ g cm^{-3} and $R_0 = 1.8 \cdot 10^8$ cm) at the edge of the star (Fig. 10.5).

10.5 Conclusion

The results of numerical modeling of the propagation of the detonation wave, with allowance for the kinetics of nuclear reactions in the CO core of a presupernova, demonstrate the evolution of the instability of the detonation front, which results eventually in its decay. We assume that the rarefaction wave, adjacent to the detonation wave, is the factor that impedes the stabilization of oscillations of the reaction zone and is responsible for the detachment of the bow shock from this zone. The decay of detonation is independent of the point of its start (in the range from the center of the CO core to its periphery). Thus, the detonation wave decays independent of the site (and, correspondingly, independent of the mechanism) of its formation, the decay time being very short. This implies that the lifetime of the detonation wave (if it appears) is determined by the time during which the rarefaction wave overtakes the shock front of the detonation wave.

In essence, we obtained a negative answer to the question posed in the title of this chapter. It is highly probable that the thermonuclear burning in the CO core develops in the *deflagration mode*. No doubt, there are factors, which were ignored in this study and which can change substantially the picture of interaction of the detonation wave with the rarefaction wave. The most evident and important among these factors might be the complicated 2D or multidimensional geometric (spin) structure of the detonation front. Such a structure can lead to the enhancement of the detonation wave, which again would leave unsolved the question of its decay.

Chapter 11

A Neutrino Transport in Type II Supernovae

Most of the energy generated in the *gravitational collapse* to a supernova is radiated in neutrino, hence the role of these particles in supernova explosion is crucial. Current models of core collapse supernovae focus on multidimensional hydrodynamics and nuclear burning, and treat neutrino transport in a simplified manner. In this chapter, an example of accurate neutrino treatment in a spherically symmetric collapse is given. The role of multidimensional effects is discussed. These results are of interest for the multidimensional models with large-scale convection as well as for the ongoing experimental search for neutrinos from supernovae.

11.1 Supernova models and neutrino

The total energy involved in the explosion of Type II SN is about 10^{53} erg, mostly released in the form of neutrinos. About 1% of this energy, namely, 10^{51} erg, is released in the form of kinetic energy of the SN ejecta. Only about 1% of that kinetic energy, i.e., 10^{49} erg, is emitted in the form of photons, which are detected as the SN event [188,282]. The relevant timescales are as follows: Collapse of the core occurs on ≤ 0.1 s timescale, the SW propagation inside the collapsing core takes ~ 10 ms, and the neutrino cooling time of the hot *neutron star* is about 10 s.

In this chapter, Type II SNe, or *core-collapse SNe*, are discussed. A progenitor of a Type II SN is a star with a zero age main sequence (*ZAMS*) *mass* exceeding 10 $M_\odot$ in which thermonuclear burning is approaching its end, yielding an iron core at its center. As the density of the matter at the center of the star becomes high enough, electrons become degenerate and relativistic, and their energy becomes sufficient for reactions in which electrons are captured by atomic nuclei, giving rise to β-unstable elements, as was first indicated in Ref. [136]. This process is called *neutronization* [61]. The first review of neutrino processes in a stellar core was given in Ref. [132].

Treatment of the explosion of core-collapse SNe can be divided into two parts with different characteristic timescales: The *gravitational collapse* and calculation of *light curves*. Light-curve calculation is the best-studied part of SN theory: Extensive observational data on visible and X-ray *light curves* and well-developed numerical models for the radiation hydrodynamics enable the determination of the explosion energy at the SN center and the chemical composition of the progenitor. At the same time, the *gravitational collapse* part remains incompletely understood. The only direct probe of the dynamics at the center of an SN is offered by neutrino. So far, few neutrino events associated with SN were observed only once [32,64,161] during the SN 1987A explosion. Overall, around 20 events were detected in a 13-s interval at a time consistent with the estimated time of the core collapse. Due to poor statistics, the determination of SN and neutrino parameters was not possible. The lack of neutrino detection [4] represents the major difficulty in understanding of the gravitational collapse and the explosion of SNe. For a recent review on predictions for neutrinos from SNe, see Ref. [226].

Models of progenitors for core collapse SNe are based on stellar evolution codes, such as KEPLER[1] or MESA[2]; for a review, see Ref. [205]. At the end of its evolution, a star with a *ZAMS mass* of 10–25 $M_\odot$ exhausts its nuclear fuel reserves, and its iron core with a mass of 1.2–1.6 $M_\odot$ begins to collapse. Neutronization initiates the photodissociation of iron nuclei and provokes instability. The collapse of a low-mass core begins at a high temperature, when

[1]https://2sn.org/kepler/doc/Introduction.html.
[2]https://docs.mesastar.org.

the mean *adiabatic index* is still $\Gamma > 4/3$, while the collapse of a massive core begins at the stability boundary $\Gamma = 4/3$ [170,177]. The collapse is accompanied by a loss of energy via neutrino emission, the *URCA process* [135,136]. The density and temperature increase during the collapse, and neutrinos carry away an energy comparable to the gravitational energy of the stellar core in its final stationary state.

Hence, one of the first *SN* explosion models are based on the *neutrino-driven mechanism* [63,109]. In this model, in order to explain the SN explosion, it is necessary to understand how less than 1% of the energy carried away by neutrinos is deposited in the envelope of collapsing star [226]. Although the problem is clear, the decade-long efforts are still far from successful [62,189]. Recent simulations [188] indicate that weak Type II SNe with kinetic energy up to $E_k \sim 10^{51}$ erg can be produced by the *neutrino-driven mechanism*, while luminous SNe with $E_k > 10^{51}$ erg and hypernovae need an alternative explanation, possibly *magneto-rotational mechanism* [66,208,249]. In the latter mechanism, rotational energy of the newly formed proto-neutron star is converted into magnetic energy. Due to large electric conductivity, magnetic fields are frozen in plasma and the role of magnetic energy increases during the collapse. Hence, the magnetic pressure or dissipative heating via *magneto-rotational instability* can drive the explosion [35,219,229,315]. Such a mechanism gives asymmetric explosions.

In what follows, the focus is on *neutrino-driven mechanism*. The first computations carried out in spherical symmetry [109] did not solve the full physical problem: The model was not correct due to the assumption that the material was transparent to neutrinos. However, the correct conclusion was reached that, provided the envelope would absorb some fraction of the neutrinos, it would be ejected [36]. Then, it was realized that the central region of the iron core is opaque to neutrinos [174,234]. Hence, the core is divided into two regions, opaque and transparent, separated by the *neutrinosphere*, a surface analogous to the photosphere in the problem of *radiative transfer*. The opacity of the neutrinos during the core collapse was first taken into account using the equations describing neutrino thermal conductivity. The limitation of such an approach is that the spectrum of neutrinos depends only on two parameters: *Chemical potential* and temperature. Besides, *neutrinosphere* radius depends on the type of neutrino and on energy. Such an approach based on thermal conductivity is relevant during the prolonged cooling of the hot newly formed *neutron stars* [154].

A more accurate treatment of neutrino transport in spherical symmetric collapse was later developed in Ref. [84]. This scheme is based on flux limiter diffusion adopted for photon transport in *neutron stars* [33]. The method for multidimensional multitemperature hydrodynamics presented in Chapter 7 of Part 2 generalizes this approach to the multidimensional case. Later, the group in the Oak Ridge National Laboratory developed an approach based on the Boltzmann equation for the spherical symmetry with all weak interactions taken into account, see Appendix B [210,220–223]. The approach adopted in Chapter 14 is similar: Opaque and transparent regions have to be considered separately. In comparison with the heat conduction models, the models with Boltzmann transport give larger average neutrino energy. It is important to stress that all spherically symmetric models of SN explosions cannot explain the energy $\gtrsim 10^{51}$ erg of expanding ejecta [62].

The reason for failure of one-dimensional models could be that convection remains unaccounted for. In the spherically symmetric problem, there are two unstable for convection regions with the negative gradient of the entropy per nucleon.[3] One region is formed for about $\gtrsim 10$ ms in the center of the forming proto-neutron star, and the second region is formed at the accretion shock. This indicates possible importance of convection for the explosion of the core collapse SN. Convection provides two effects: (1) The neutrino energy flux through the envelope can be increased; (2) the average neutrino energy is increased as well; this effect obviously requires a multidimensional approach.

Multidimensional models based on neutrino diffusion with the flux limiters or simplified transport along the radius were developed in Refs. [120,231,288]. Simulations in 2D [121,232] and 3D [111,318] apparently successfully generate *large-scale convection* leading to explosion. Careful examination of the resolution effects in 3D simulations show that the dynamics of the turbulent cascade of energy

[3]Generally speaking, the Ledoux criterion [209] should be used due to the conservation of leptonic number [71].

from large to small scale is severely affected by numerical viscosity [112,263] and hence successful explosions in 3D models might be numerical artifact.

As summarized above, 1D models do not account for turbulence and convection and other multidimensional effects, such as *standing accretion shock instability* [78] and the *lepton-emission self-sustained asymmetry* [289], and do not show explosions in a self-consistent way. Nevertheless, spherically symmetric simulations are a very efficient way for the detailed study of the neutrino luminosities and *spectra*, in particular for the long-time cooling evolution of the newly formed *neutron star* over tens of seconds. Accurate hydrodynamic simulations [112,263] indicate that convection develops on small scales only, hence such study is relevant for observations of neutrino signal from core collapse SNe. Therefore, even the development of 1D models that include sufficiently complete physics remains very topical [211,223,246].

In this chapter, spherically symmetric collapse with full account of neutrino interactions is considered using the finite difference approach described in Chapter 6 of Part 2.

11.2 1D problem and initial model

As recalled in the previous section, evolution of massive stars with masses in the range from 15 $M_{\odot}$ to 25 $M_{\odot}$ ends with the formation of iron cores with masses 1.2–1.6 $M_{\odot}$ [245,313]. In this section, *gravitational collapse* of an iron core with mass 1.4 $M_{\odot}$ is considered.

The computations include the solution of the hydrodynamic equations for the matter interacting electromagnetically. The transport kinetic Boltzmann equations are applied for the neutrinos and also the kinetics equations for the difference between the numbers of electrons and positrons per nucleon Y_e [7] are included. The kinetics of neutrino production and capture by free nucleons and nuclei, as well as neutrino scattering on electrons and positrons and other weak interactions are taken into account. All these interactions conserve lepton number. On the one hand, only two-particle interactions are treated, while, on the other, there are no numerical methods which conserve the lepton number exactly. Also, degeneracy of electrons should be taken into account.

The hydrodynamics and transport equations are solved in spherical symmetry in a *comoving reference frame*, with terms of first order in v/c in a Newtonian gravity.

The *method of lines* is used for the numerical integration of hydrodynamic and transport equations. The thermodynamic functions of the electron–positron gas are computed using the code described in Ref. [75]. In comparison with previous treatments [211,220,222], the adopted scheme is fully implicit and uses high-order *Gear's method* with the automatic selection of the order from 1 to 5 in order to optimize time steps. In addition, unlike the treatment in Ref. [221], the described approach is based on the exact evaluation of the matrix elements for the neutrino scattering on electrons. Neutrino annihilation is computed assuming isotropic distribution of both neutrinos and anti-neutrinos.

In the Lagrangian variables $m = \int_0^r (r')^2 \rho(r',t)dr'$ and t, the hydrodynamic equations can be written in the form (4.1)–(4.5).

In order to close the system (4.1)–(4.4), one should have an expression for the emission and absorption coefficients on the right-hand side of (4.4) and should specify the *equation of state*, $P = P(\rho, T, Y_e)$.

A computational grid in the *phase space* is introduced, $(\epsilon_\omega, \mu_k, r_j)$. The radii $r_{j+1/2}$ and velocities $v_j + 1/2$ are determined at the cell boundaries, while the specific energy of the matter ϵ_j, number of electrons per nucleon $Y_{e,j}$, and *spectral energy density* of the neutrinos $E_{\nu,j,k,\omega}$ are determined as the mean values within the phase volume. The spatial derivatives (4.1)–(4.3) are approximated using the *central differences* [7] with *artificial viscosity.* Introducing the notation $\Delta(f)_j \equiv f_{j+1/2} - f_{j-1/2}$ and the analogous notation for the subscript k, the spatial derivatives in the Boltzmann equation for (4.4) can be approximated [7]:

$$\frac{1}{c}\frac{\partial \Delta E_{\nu,j,k,\omega}}{\partial t} + \frac{\Delta(r^2 \bar{\mu}_k \Delta E_{\nu,k,\omega})_j}{\Delta V_j}$$
$$+ \left(\frac{\bar{1}}{r}\right)_j \frac{1}{\Delta \mu_k}\left((1-\mu^2)\left[1 + \left(\frac{3\bar{v}_j}{2c} - \frac{\bar{r}_j}{2c}\frac{\Delta(r^2 v)_j}{\Delta V_j}\right)\mu\right]\Delta E_{\nu,j,\omega}\right)_k$$
$$+ \left(\frac{\bar{1}}{r}\right)_j \left[\frac{\Delta\mu_k^3/3}{\Delta\mu_k}\left(\frac{3\bar{v}_j}{c} - \frac{\bar{r}_j}{c}\frac{\Delta(r^2 v)_j}{\Delta V_j}\right) - \frac{\bar{v}_j}{c}\right]\frac{\Delta(\epsilon_\nu \Delta E_{\nu,j})_\omega}{\Delta \epsilon_{\nu,\omega}}$$

$$+\left(\overline{\frac{1}{r}}\right)_j \left\{\frac{\bar{v}_j}{c} + \frac{\bar{r}_j}{c}\frac{\Delta(r^2v)_j}{\Delta V_j} - \frac{\Delta\mu_k^2/2}{\Delta\mu_k}\left(\frac{3\bar{v}_j}{c} - \frac{\bar{r}_j}{c}\frac{\Delta(r^2v)_j}{\Delta V_j}\right)\right\}$$

$$\times\Delta E_{\nu,j,k,\omega}$$

$$= -\chi_{\nu,j,\omega}\Delta E_{\nu,j,k,\omega} + \Delta\eta_{\nu,j,\omega}, \tag{11.1}$$

where $\Delta E_{\nu,j,k,\omega} = \Delta\epsilon_{\nu,\omega}E_{\nu,j,k,\omega}$, $\left(\overline{1/r}\right)_j = \frac{\Delta r_j^2/2}{\Delta r_j^3/3}$, $\bar{r}_j = (r_{j-1/2} + r_{j+1/2})/2$, $\bar{v}_j = (v_{j-1/2} + v_{j+1/2})/2$, and $\bar{\mu}_k = (\mu_{k-1/2} + \mu_{k+1/2})/2$.

The approximation for the flux along the direction r can be written as

$$(\bar{\mu}_k\Delta E_{\nu,k,\omega})_{j+\frac{1}{2}}$$

$$= \left(1 - \eta_{\nu,j+\frac{1}{2},\omega}\right)\left(\frac{\bar{\mu}_k + 1}{2}\Delta E_{\nu,j,k} + \frac{\bar{\mu}_k - 1}{2}\Delta E_{\nu,j+1,k}\right)$$

$$+\eta_{\nu,j+\frac{1}{2},\omega}\frac{\Delta E_{\nu,j,k} + \Delta E_{\nu,j+1,k}}{2}, \tag{11.2}$$

$$\eta_{\nu,j+\frac{1}{2},\omega} = \frac{\chi_{\nu,j,\omega}\Delta r_j\chi_{\nu,j+1,\omega}\Delta r_{j+1}}{\chi_{\nu,j,\omega}\Delta r_j + \chi_{\nu,j+1,\omega}\Delta r_{j+1}} \Bigg/ \left(1 + \frac{\chi_{\nu,j,\omega}\Delta r_j k_{\nu,j+1,\omega}\Delta r_{j+1}}{\chi_{\nu,j,\omega}\Delta r_j + \chi_{\nu,j+1,\omega}\Delta r_{j+1}}\right). \tag{11.3}$$

The approximation for the flux along the direction μ using a second-order approximation and *integrating along the characteristic* can be written as

$$\Delta E_{\nu,j,k+\frac{1}{2},\omega} = \begin{cases} \Delta E_{\nu,j,k,\omega} + \frac{\Delta\mu_k(\Delta E_{\nu,j,k,\omega} - \Delta E_{\nu,j,k-1,\omega})}{(\Delta\mu_k + \Delta\mu_{k-1})}, \\ \text{for } 1 + \left(\frac{3\bar{v}_j}{2c} - \frac{\bar{r}_j}{2c}\frac{\Delta(r^2v)_j}{\Delta V_j}\right)\mu_k \geq 0, \\ \Delta E_{\nu,j,k+1,\omega} - \frac{\Delta\mu_{k+1}(\Delta E_{\nu,j,k+2,\omega} - \Delta E_{\nu,j,k+1,\omega})}{(\Delta\mu_{k+1} + \Delta\mu_{k+2})}, \\ \text{for } 1 + \left(\frac{3\bar{v}_j}{2c} - \frac{\bar{r}_j}{2c}\frac{\Delta(r^2v)_j}{\Delta V_j}\right)\mu_k < 0. \end{cases} \tag{11.4}$$

With these expressions for $(\bar{\mu}_k\Delta E_{\nu,k,\omega})_{j+\frac{1}{2}}$ $\Delta E_{\nu,j,k+\frac{1}{2},\omega}$, in the opaque region, where $\chi\Delta r_j$ is small and the transport dominates, the spatial derivatives in (11.1) against the flow can be approximated in accordance with the direction of propagation of the perturbations. False oscillations are absent in the numerical solution, but the scheme is first order in the derivatives in r. In the opaque region, the local energy exchange between the neutrinos and matter dominates, and approximating the derivatives in r using the *central differences* does not lead to false oscillations. The resulting second-order approximation in the spatial derivatives is important for the correct behavior of the numerical solution in the region with high *optical depth.* There is a transition to the equations of neutrino thermal conductivity. A similar approximation for the spatial derivatives in the equation for the neutrino distribution function *occupation number* is adopted in Ref. [220].

The system of ODEs

$$\dot{\mathbf{y}} = f_i(\mathbf{y}) \tag{11.5}$$

for the unknowns $\mathbf{y} = r_{j+1/2}, v_{j+1/2}, \epsilon_j, \Delta E_{\nu,j,k,\omega}, Y_{e,j}$ is solved numerically. Since the rate at which matter flows through and the rate at which neutrinos are redistributed can differ substantially, the eigenvalues of the *Jacobi matrix* can differ appreciably. Hence, *Gear's method* is used again.

The *equation of state* takes into account the equilibrium radiation of photons, the electron–positron gas, and a mixture of nuclei in the so-called *nuclear statistical equilibrium* with free nucleons [279]:

$$(A_i, Z_i) \rightleftarrows (A_i - Z_i)n + Zp. \tag{11.6}$$

The neutrons and protons have statistical weights $\omega_{n,p} = 2$. The following nuclei are considered: ${}_2\mathrm{He}^4$ nuclei with $\omega_{\mathrm{He}} = 1$ and binding energy per nucleon $Q_{\mathrm{He}} = 28.296$ MeV and ${}_{26}\mathrm{Fe}^{56}$ nuclei with $\omega_{\mathrm{Fe}} = 1$ and $Q_{\mathrm{Fe}} = 492.322$ MeV, where $Q_i = c^2\left[Z_i m_p + (A_i - Z_i)m_n - m_i\right]$. *Fermi–Dirac statistics* is assumed for the nucleons in the *equation of state* in a non-relativistic approximation; for details, see Ref. [14].

The dependence of the *pressure* on the density in the initial state is a polytropic *equation of state* $P = P(\rho) = K\rho^{1+1/n}$, with the polytropic index $n = 3$ corresponding to the *pressure* of a degenerate, relativistic electron gas.

The central density ρ_c was chosen so that the star was near the boundary of *instability to collapse* (more precisely, the approximate boundary of the region of stability):

$$\langle \Gamma \rangle \equiv \int \frac{dm}{\rho} \Gamma P \bigg/ \int \frac{dm}{\rho} P = \frac{4}{3}, \tag{11.7}$$

where $\Gamma = \left(\frac{\partial \ln P}{\partial \ln \rho} \right)_s$ is the *adiabatic index*. The collapse of a massive stellar core $\gtrsim 2\ M_\odot$ occurs due to the photodissociation of iron nuclei precisely at the stability boundary, $\langle \Gamma \rangle = 4/3$. For less massive stars with iron-core masses $\approx 1.4\ M_\odot$, the collapse occurs at higher central densities and temperatures, when the neutrino energy losses during the *neutronization* of the core becomes important, but $\langle \Gamma \rangle > 4/3$.

The global parameters for initial state are presented in Table 11.1.

The $100 \times 8 \times 12$ grid for the mass coordinate $m = \int_0^r (r')^2 \rho(r', t) dr'$, the cosine of the angle between the neutrino velocity and the radius vector μ, and the energies of the neutrinos and anti-neutrinos ϵ_ν, $\epsilon_{\tilde{\nu}}$ is used for the computations of the collapse. The numerical results are summarized in Table 11.1. The collapse begins after roughly 4 s, when the density and temperature at the center of the star and the kinetic energy grow sharply. The stellar radius takes on its minimum value 5×10^7 cm, at time 4.2 s, see Fig. 11.2. The time for the transition to the collapse depends on the energy losses to neutrino emission and is determined by the initial density and temperature. This time is much longer than the characteristic timescale for the initial state, which is $\simeq 1/\sqrt{GM/R_s^3} = 0.8$ s. At the initial times, the energy of the neutrino emission is modest compared to the energy of the matter; the power released in the form of neutrinos and anti-neutrinos is $-dE_t/dt \sim 10^{48}$ erg s^{-1}, see the neutrinos luminosities in Fig. 11.1. Due to the slow rate at which the parameters of the matter change, the neutrino energy flux, which is

Table 11.1 Characteristics of the stellar core at the beginning and end of the computations.

Time (s)	$E_{\rm total}$ (erg)	$E_{\rm gr}$ (erg)	$E_{\rm kin}$(erg)	ρ_c (g cm$^{-3)}$	T_c (K)	$R_{\rm star}$ (cm)
$t = 0$	-5.1×10^{50}	-3.7×10^{51}	0	4.0×10^9	7.2×10^9	1.9×10^8
$t = 40$	-8.3×10^{52}	-1.8×10^{53}	3.4×10^{50}	2.8×10^{14}	5.2×10^{10}	2.7×10^6
						$M = 1.353\ M_\odot$

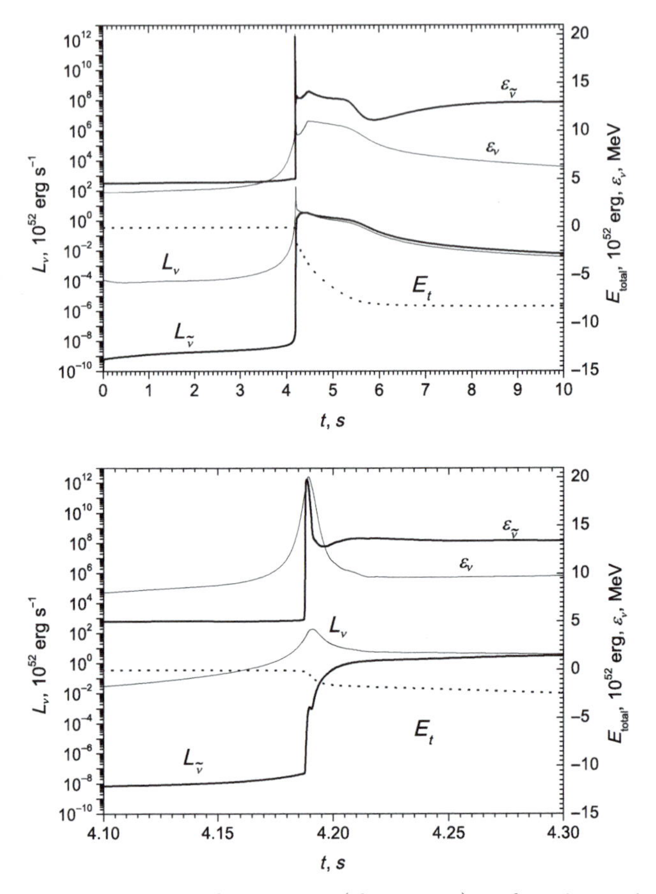

Figure 11.1 Luminosity of neutrinos (thin curve) and anti-neutrinos (thick curve), mean energy of neutrinos (thin curve) and anti-neutrinos (thick curve), and the total energy of the star (dotted curve) at various times. Calculations confirm the results in Ref. [14], Fig. 4.

primarily emitted from the inner regions of the star, rapidly reaches the stellar surface and is a constant along the mass coordinate m for the outer layers of the star.

With increase in the density and temperature at the center of the star (ρ_c, T_c), the flux of energy carried away by neutrinos grows sharply, accelerating the collapse. However, when the value $\rho_c \sim 10^{11}$ g/cm^3 is reached, a neutrino-opaque region appears near the center, bounded by the *neutrinosphere*. In this opaque region inside the neutrinosphere, the neutrino transport is slowed by absorption.

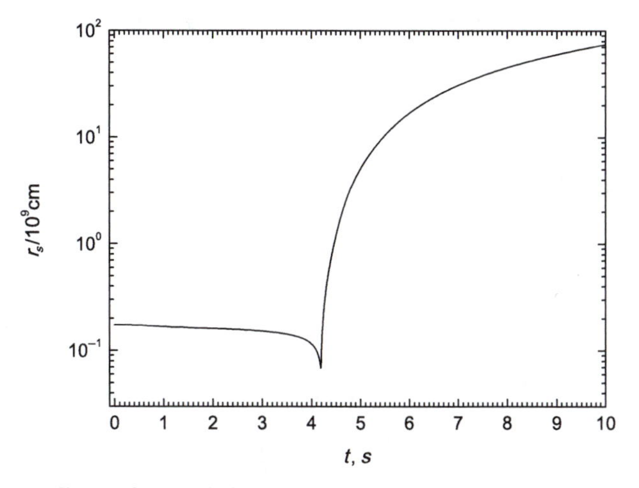

Figure 11.2 Dependence of the star radius on time. Calculations confirm the results in Ref. [14], Fig. 3.

Matter is accreted from outside, giving rise to a shock. The transition to collapse occurs very sharply: The central density increases from 3×10^{10} g/cm^3 to 2×10^{11} g/cm^3 (when the *neutrinosphere* forms) over only 100 ms and reaches values 10^{13} g/cm^3 after only another 15 ms. This time is comparable to the time for a neutrino to travel from the center to the surface of the star, $R_s/c = 18$ ms (during which time the stellar radius changes only slightly). See also Fig. 11.2.

At approximately $t_1 = 4.19$ s, the maximum energy flux in neutrinos is reached at the *neutrinosphere*. At later times, the flux from it is lower, and a growth of the flux is observed outward from the *neutrinosphere*, which then decreases toward the stellar surface. This non-monotonic behavior is associated with the finite velocity of the neutrinos as well as the motion and compression of the matter. The maximum energy flux at the *neutrinosphere* is reached at the time when the shock has its maximum intensity, see Fig. 11.3. A large growth in flux near it by a factor of a few is accompanied by strongly non-equilibrium kinetics. The degree of *neutronization* Y_e in the accreting flow decreases by nearly a factor of two in the transition through the shock inside the *neutrinosphere*, see Fig. 11.4. This is *neutronization* on the *SW*. A maximum of $Y_e(m)$ is observed in front of the shock, with $Y_e > 0.47$, obviously due to the absorption

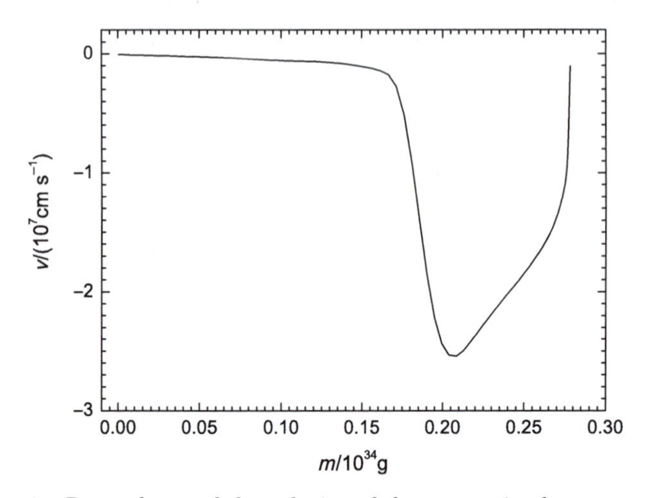

Figure 11.3 Dependence of the velocity of the matter in the star on the mass coordinate at time 4.19 s. Calculations confirm the results in Ref. [14], Fig. 5.

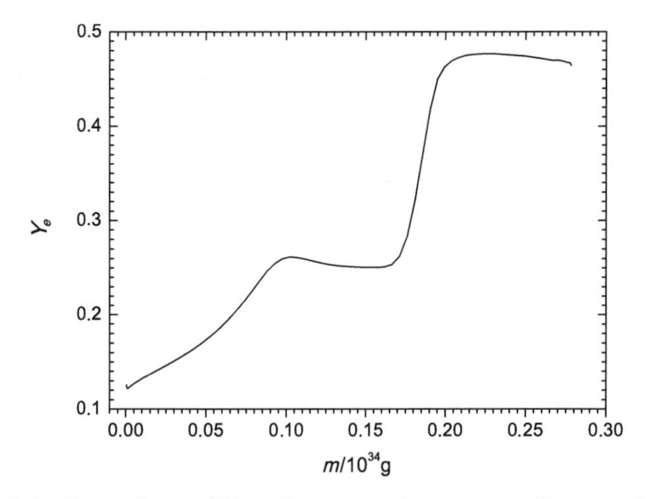

Figure 11.4 Dependence of Ye in the star on the mass coordinate at time 4.19 s. Calculations confirm the results in Ref. [14], Fig. 6.

of neutrinos by free neutrons, which form due to the dissociation of iron at the enhanced temperature.

The time dependence of *neutrino luminosity* $L(t)$, the total energy $E(t) = E(0) - \int_0^t L(t')dt'$, and the mean energy of the neutrinos are presented in Fig. 11.1. The total energy radiated over the entire collapse is 1.2×10^{53} erg. The light curve maximum due to neutrinos reaches 1.8×10^{54} erg s^{-1} at time $t = 4.195$ s and has a half-width of 10 ms. During this time, the total energy of the star is decreased by $\approx 8 \times 10^{51}$ erg, which comprises a few percent of the total radiated energy, and the kinetic energy is strongly reduced (by several orders of magnitude). The maximum kinetic energy, 2.9×10^{51} erg, is reached at time $t = 4.188$ ms. This is 10 ms earlier than the neutrino luminosity maximum, which corresponds to the time for neutrinos to travel from the *neutrinosphere* to the stellar surface. The appearance of a narrow maximum in the luminosity is due to the reprocessing of kinetic energy and the release of the gravitational energy of the infalling flow of matter in the region of the *neutrinosphere* and shock. The characteristic mean energies of the particles are 20 MeV in the region of the radiation maximum and 10–17 MeV at later times for neutrinos and anti-neutrinos, respectively.

Narrow maxima with values $(2\text{–}6) \times 10^{53}$ erg s^{-1} in neutrino luminosities with characteristic widths of tens of milliseconds have not appeared in the solutions obtained in other studies [85,233] as is true in the present work roughly at the time of formation of the shock in the star. Time-resolved measurements of this narrow maximum in the neutrino luminosity could be used to estimate the mass of the neutrinos [324]. The relationship between energy and velocity for particles with non-zero mass is given by $\epsilon_\nu = \frac{m_\nu c^2}{\sqrt{1-v^2/c^2}}$. For high energies and ultra-relativistic velocities, the delay time for the arrival of the neutrinos is

$$\tau_1 = \frac{1}{2}\left(\frac{m_\nu c^2}{\epsilon_\nu}\right)^2 t_0, \tag{11.8}$$

where t_0 is the light-travel time from the star to the observer. For a collapse in our galaxy (at a distance of 10 kpc) $t_0 = 1.0 \times 10^{12}$ s. Due to the difference in the arrival times, an observer on Earth sees the neutrino luminosity peak to be spread out in time:

$$L_{\mathrm{E}}(\tau) = \int_0^\infty dt' \left.\frac{dL}{d\epsilon_\nu}\right|_{\tau - t' + t_1} \frac{\epsilon_\nu}{2(\tau - t')}, \tag{11.9}$$

where $\epsilon_\nu = m_\nu c^2 \sqrt{\dfrac{t_0}{2(\tau - t')}}$. If the neutrino peak were observed and found to have a width ≈ 10 ms, this would constrain the *neutrino mass* to be $\lesssim 4$ eV: At higher masses, the broadening of the *light curve* would become appreciable for a source at a distance of 10 kpc, i.e., a source in our galaxy. This constraint on the *neutrino mass* is of little practical interest. However, according to Eq. (11.8), the delay is proportional to the distance to the observer and inversely proportional to the square of the particle energy. In other words, for a collapse at a distance a factor of k^2 larger than our previously chosen 10 kpc, or with particle energies a factor of k lower than the mean energy at the maximum, it follows a constraint on the *neutrino mass* that is a factor of k more stringent. High energies ($\gtrsim 10$ MeV) are favorable for detection with neutrino observatories, such as Super-Kamiokande.

The second notable result is the absorption of part of the energy of the neutrinos and anti-neutrinos in outer layers of the stellar core, see Table 11.1. This same effect was manifest in the computations [8,28] and increases with decreasing core mass. This effect is believed to be due to inaccuracies in the initial model, the simplified *equation of state* used, and the assumption of spherical symmetry. However, it is important to bear this effect in mind in further studies of SN mechanisms. First, specially conducted computations with a zero neutrino-absorption coefficient do not give rise to expansion of the envelope, i.e., the obtained weak explosion is associated with neutrino absorption, not a hydrodynamical outflow. Second, neutrinos have substantially higher energies compared to the case where the approximation of neutrino thermal conductivity is adopted. This effect can occur in multidimensional computations with *large-scale convection*, when neutrinos from deep layers with higher energies rise toward the surface due to large-scale motions in the region of convection. The *cross-section* for scattering on electrons increases with increasing neutrino energy [203]

$$\sigma_{e\nu} = 1.7 \times 10^{-40}\ \mathrm{cm}^2 \frac{\epsilon_\nu}{m_e c^2}\frac{E_F}{m_e c^2}, \quad \text{for } E_F \ll \epsilon_\nu, \tag{11.10}$$

where the *Fermi energy* of the electrons is $E_F = (3\pi^2)^{1/3}\hbar c n_e^{1/3}$. According to (B.5) in a collision with a stationary electron, $\epsilon_e = E_F$, the neutrino transfers a fraction of its energy

$$\frac{\epsilon_\nu - \epsilon'_\nu}{\epsilon_\nu} = \frac{\epsilon_\nu}{E_F + \epsilon_\nu} \approx 1 \text{ for } E_F \ll \epsilon_\nu. \tag{11.11}$$

Thus, a shell with thickness l_{sh} and with a density of degenerate electrons n_{sh} absorbs a fraction of the neutrino energy

$$\frac{E_{\text{k}}}{E_{\text{total}}(t=0) - E_{\text{total}}(t_{\text{fin}})} = \sigma_{e\nu} l_{\text{sh}} n_{\text{sh}}. \quad (11.12)$$

For an estimate, take $l_{\text{sh}} = R_{\text{star,min}} = R_{\text{star}}(t = 4.3648 \text{ s}) = 5 \times 10^7$ cm (Fig. 11.2), and to estimate the number density, take the mean density for the ejected mass 0.047 $M_\odot$ from Table 11.1. Thus, the number of electrons in the shell is $N_{\text{sh}} = 2.8 \times 10^{55}$, $n_e = 5.3 \times 10^{31}$ cm^{-3}, $E_F/(m_e c^2) = 4.5$. For an absorbed fraction of the neutrino energy $E_{\text{k}}/(E_{\text{total}}(t=0) - E_{\text{total}}(t_{\text{fin}})) = 0.0041$, one should obtain a mean energy for the neutrinos of ~ 10 MeV, in agreement with the computations, i.e., the ejection of the envelope is associated with the transfer of some part of the neutrino energy.

At the end of the computations, a stationary, cool, degenerate *neutron star* is obtained. Kinetic energy of the neutron star is close to zero, the difference of the temperature from zero only insignificantly influences *pressure* and energy. On account of the inclusion of the gas *pressure* due to interacting baryons in the equation of state, the *equation of state* for the final models of the star are not polytropic. As the temperature decreases, general relativistic effects, neglected here, become dominant, and also a more realistic equation of state for nuclear densities is required.

11.3 Multidimensional models of SN and large-scale convection scales

The *large-scale instability* (*convection*) of a hot, collapsing, proto-neutron-star core was first considered in Ref. [101]. The *equation of state* included the relativistic electron–positron gas and an ideal ion gas. An equilibrium configuration with a central density of 2×10^{13} g cm^{-3} was adopted for the initial state. A constant specific entropy was chosen, except for the central region, where a high temperature and entropy were established due to *neutronization* behind the accretion shock. 3D modeling indicated the development of convection, manifest as the upward rising of a hot bubble. Convection can arise when $\mathbf{g}\nabla s > 0$.

Here, the study connected with *large-scale convection* considering the evolution of the *neutrino spectrum* in a hot bubble [287] is reported. The neutrinos were initially assumed to be captured in the hot bubble. The evolution of their spectrum was studied until the expansion of the homogeneous bubble [101], taking into account the density variations with time in accordance with the solution obtained in Ref. [101]. Therefore, there was no dependence on spatial variables or angles in the model, the energy and time dependence of the neutrino DF was retained in the homogeneous and isotropic treatment, and emission, absorption, and scattering reactions with the participation of electrons and nuclei were taken into account. This yielded an important result: The mean neutrino energy decreased to $\gtrsim 40$ MeV during the expansion. This implies that the means used to redistribute the *neutrino spectrum* to higher energies adopted in the previous section are completely justified.

The interactions between high-energy neutrinos and the envelope of a collapsing stellar core were also taken into account in Ref. [53], where the Euler gas dynamics of the *SN* envelope was considered, allowing for neutrino absorption. The dependence of the neutrino emission on the coordinates (r, θ) during the development of convective instability was specified. The total *neutrino luminosity* was assumed to be constant 5×10^{52} erg s^{-1}, and the neutrino energy spanned in the range 30–60 MeV. The kinetic energy of the expanding envelope was determined to be $(1.5–50) \times 10^{51}$ erg, depending on the initial accretion rate of the envelope material.

In Fig. 11.5, the entropy profiles are shown at different time moments. On can see two unstable regions $ds/dr < 0$: One region near the center during $\gtrsim 10$ ms, and another region on the accretion SW during long time interval. More accurate criterion for the convection, called *Ledoux criterion*, is [209]

$$\left(\frac{\partial P}{\partial s}\right)_{\rho Y_l} \frac{ds}{dr} + \left(\frac{\partial P}{\partial Y_l}\right)_{\rho s} \frac{dY_l}{dr} < 0, \tag{11.13}$$

where l is the number of leptons. In the central region inside the *neutrinosphere*, the neutrinos are trapped, so approximate *Schwarzschild criterion* for specific entropy is appropriate. See also Fig. 11.6.

We consider a multicomponent gas of different substances i described by a set of concentrations n_i, $\rho\epsilon_\alpha(\mathbf{r}, t)$ are internal energy densities, ϵ_α is specific energy, see Chapter 7 of Part 2 and [11,29,30,

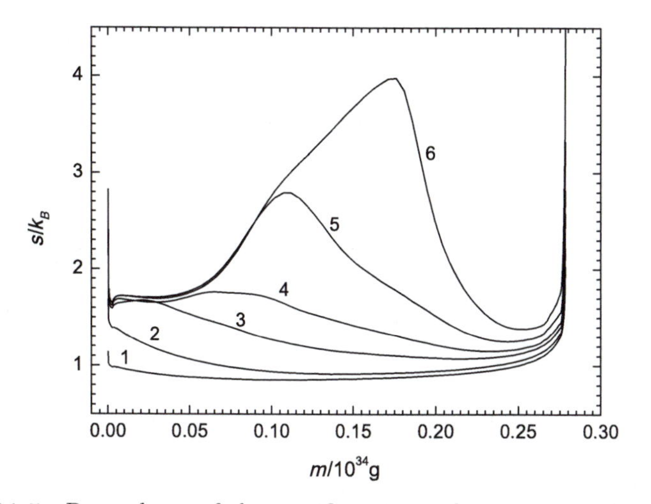

Figure 11.5 Dependence of the specific *entropy density* per nucleon s/k_B in the star on the mass coordinate at times $t = 4.1759$ (1), 4.18360 (2), 4.1866 (3), 4.1873 (4), 4.1882 (5), and 4.1893 (6) s. The times are denoted by the numbers near the curves. The entropy in the first mass interval, $m \approx 0$, exceeds the real physical value and it is the numerical artifact.

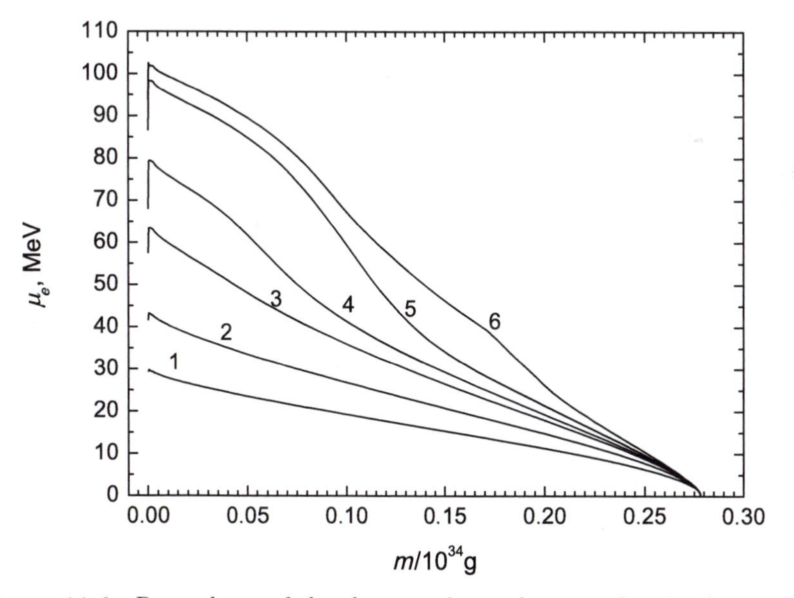

Figure 11.6 Dependence of the electron *chemical potential* μ_e in the star on the mass coordinate at the same times as those in Fig. 11.5.

121,309]. All massive particles (nuclei, nucleons, electrons, positrons) have identical velocities $\mathbf{v}(\mathbf{r}, t)$ and temperatures, while the neutrinos have their spectral energy densities $\rho\epsilon_{\nu,i}$, can transfer energy by heat conduction not associated with the velocity of the massive particles, and can participate in reactions. The basic mathematical problem concerns the gas-dynamics part because of the discontinuities in the solution. We carry out the hydrodynamic transport by using the Godunov-type method based on the Riemann problem solver for the mixture. In a joint treatment of the matter and fast particles, the transport only weakly affects the thermal distribution, and reactions computed using an implicit scheme in individual steps do not influence the number of time steps required for the gas-dynamics transport.

We have a system of Euler equations that can be used to describe the gravitational collapse. The equation for the number density of baryons is

$$\frac{\partial \rho/m_p}{\partial t} + \mathrm{div}((\rho/m_p)\mathbf{v}) = 0, \tag{11.14}$$

and the equation for the difference between the concentrations of electrons and positrons is

$$\frac{\partial \Delta n_e}{\partial t} + \mathrm{div}(\Delta n_e \mathbf{v}) = \dot{Y}_e \rho/m_p. \tag{11.15}$$

The velocities of all components with non-zero mass were taken to be equal, with energy transport occurring for massless particles. The conservation of momentum for the matter can be written as

$$\frac{\partial \rho v_j}{\partial t} + \nabla_i \Pi^{\mathrm{m}}_{ij} = \rho g_j + \rho f_\nu, \tag{11.16}$$

and the equation for the energy density of the matter is

$$\frac{\partial \rho E_{\mathrm{m}}}{\partial t} + \mathrm{div}(E_{\mathrm{m}}\rho + P_{\mathrm{m}})\mathbf{v} = \rho \mathbf{v}\mathbf{g} + \rho q_{\mathrm{m}}. \tag{11.17}$$

The term ρq_m represents heating of the matter by neutrinos. The gravity acceleration is the gradient of the potential $\mathbf{g} = -\nabla\Phi$, obtained from the Poisson equation. Poisson equation solver is described in Chapter 9 of Part 2 and in Ref. [10].

In the non-relativistic assumption $v \ll c$, the distribution functions in the laboratory frame and a frame comoving with the matter are the same. The neutrino spectrum in thermal equilibrium described by the equilibrium temperature and the neutrino chemical potential equals the chemical potential of electrons $\mu_\nu = \mu_e$, $f_\nu^{\rm eq} = \frac{2}{(2\pi\hbar c)^3}\left(1+\exp\left(\frac{\varepsilon-\mu_\nu}{kT^{\rm eq}}\right)\right)^{-1}$. We must introduce a grid for the neutrino energy and use the spectral energy density of the neutrinos and anti-neutrinos, $\rho\Delta\varepsilon_{\nu,\omega} \equiv \int_{\varepsilon_{\omega-1/2}}^{\varepsilon_{\omega+1/2}} d\varepsilon\varepsilon_\nu(4\pi c^3\varepsilon_\nu^2 f_\nu)$, in each grid interval $(\varepsilon_{\omega-1/2}, \varepsilon_{\omega+1/2})$.

The equation of transport for the spectral energy density of neutrino is

$$\frac{\partial\rho\Delta\varepsilon_{\nu,\omega}}{\partial t} + \mathbf{v}\nabla\left(\rho\Delta\varepsilon_{\nu,\omega}\right) = \mathrm{div}\Delta\mathbf{F}_{\nu,\omega} - \rho\Delta q_{\mathrm{m},\omega}, \tag{11.18}$$

where the flux is determined by the gradient of the zeroth moment. In the opaque case, $\Delta\mathbf{F}_{\nu,\omega}^{\rm thick} = -\frac{1}{3\chi}\mathrm{grad}\rho\Delta\epsilon_{\nu,\omega}$, and in the transparent case, $\Delta F_{\nu,\omega}^{\rm max} = c\rho\Delta\epsilon_{\nu,\omega}$. In the arbitrary case, we can use the interpolation (the *flux limiter diffusion*), $\mathbf{\Delta F}_{\nu,\omega} = \frac{\Delta\mathbf{F}_{\nu,\omega}^{\rm thick}}{\left|\Delta\mathbf{F}_{\nu,\omega}^{\rm thick}\right|/\Delta F_{\nu,\omega}^{\rm max}+1}$. The nonlinear diffusion flux transfers a parabolic equation in the opaque region into hyperbolic transport equation in the transparent region for the spectral energy densities.

The exchange of energy between neutrinos and matter is described by the relaxation to a thermal distribution, $\rho\Delta q_{\mathrm{m},\omega} = c\chi\left(\rho\Delta\epsilon_{\nu,\omega} - \rho\Delta\epsilon_{\nu,\omega}^{\rm th}\right)$. The relaxation rate was chosen to be proportional to the concentration of free nucleons, $\chi \approx \sigma_0 n_n$, with the constant cross-section of weak interactions $\sigma_0 = 1.7 \times 10^{-40}$ cm^2. We adopted the relaxation time for the difference of the numbers of electrons and positrons per nucleon ($Y_e \equiv \Delta n_e m_p/\rho$) $\tau_0 = 10^{-3}$ s $\frac{10^{12}\ \mathrm{g\ cm^{-3}}}{\rho}\left(5\times 10^{10}\ \mathrm{K}/T^{\rm eq}\right)^3$ and the relaxation rate $\dot{Y}_e = -(Y_e - Y_e^{\rm eq})/\tau_0$ toward the *beta-equilibrium distribution* from [68], where only two reactions were considered: $e^- + p \to n + \nu$ and $e^+ + n \to p + \tilde{\nu}$, with the free escape of the neutrinos. Such beta equilibrium is correct only near the neutrinosphere, where it is important to take the energy flux of the neutrinos into account. In a deeply opaque region, the simple beta-equilibrium model is quantitatively incorrect, however, the neutrino energy flux is small in the opaque region.

The EOS of the matter, $P_m = P_m(\rho, \epsilon_m, Y_e)$, contains the free nucleons and the nuclei in statistical equilibrium. Also, matter contains degenerate ultra-relativistic pairs and photons. As initial data, we adopted the relationship between the pressure and density for a polytrope, $P \propto \rho^{1+1/n}$, with the polytropic index $n = 3$, corresponding to ultra-relativistic electrons. The rotation law is a constant ratio of the centrifugal and gravitational forces in the equatorial plane (it is a generalization of rigid-body rotation for an incompressible fluid), see Chapter 8 of Part 2 and Ref. [12]. With modest initial rotation, the ratio of the polar radius to the equatorial radius r_{eq} was taken to be 0.9, and the rotational energy was 1.25% of the gravitational energy.

The chosen law of *differential rotation* (the angular velocity depends on the cylindrical radius) makes it possible to consider arbitrarily large rotational energies, up to a thin disk, attainable only at the end of compression, when after energy losses due to neutrino radiation, the gravity is mainly held by the centrifugal force. On the other hand, at the beginning of gravitational collapse for the core of a star with radius of $\sim 10^9$ cm, we should have chosen a rigid-body rotation of polytrope $n = 3$, since the combustion processes in a massive star are accompanied by convection, and the angular velocity is equalized. The spatial distribution of density in polytrope turns out to be rather loose, with the strongest rigid-body rotation, for it is obtained when the first cosmic velocity with the rotation energy and angular momentum erg s is reached at the equator [13]. Therefore, the selected differential rotation in the formulation of the problem can be considered weak relative to the maximum possible event at the stage of momentum collapse. For a loose configuration of polytrope with rigid-body rotation, the role of centrifugal force is large toward the edge of the star; in the case of the differential law of rotation adopted at the beginning of calculations, the ratio of centrifugal force to gravity in the entire equatorial plane is constant exactly as for a rigid equation of state for an incompressible fluid. Rotation effects affect the central region where convection is of interest.

The polytropic initial model contains three independent physical parameters: The gravitational constant G, central density ρ_c, and equatorial radius r_{eq}. For the central density $\rho_c = 2 \times 10^{12}\text{g/cm}^3$ and the mass of the collapsing core of 1.4 $M_\odot$ adopted for calculations, the polytropic model gives the equatorial radius $r_{\rm eq} = 2.68 \times 10^7$ cm, gravitational energy $E_{\rm gr} = 2.93 \times 10^{52}$ erg, and angular momentum $J_z = 9.87 \times 10^{48}$ erg s. For the specified density and pressure profiles, we must calculate the parameters T and $Y_e = Y^{\rm eq}$ from the temperature $T = T(\rho, \epsilon = 3P/\rho)$ corresponding to the initial equilibrium (without taking into account neutrino energy losses) at time $t = 0$. The specified central density in the initial equilibrium state corresponds to the collapse of a real stellar core that is already underway. The small initial radius makes it possible to resolve the formation of a neutron star with a radius of ~ 10 km on a fixed Euler computational grid. In reality, the collapse begins at a stellar radius of $\sim 10^8$ cm, but the initial stage of the neutrino energy losses continues for several seconds, appreciably exceeding the gas-dynamical timescale $(G\rho)^{-1/2}$, even for a low initial central density $\sim 10^9$ g cm^{-3} [14]. As a result of reconstructing the solution in the 2D problem, we obtained an entropy profile that falls off toward the center, $ds/dr < 0$. This type of profile arises due to neutronization, which lowers Y_e, so that the specific energy from the electron component goes into nucleons (Fig. 11.7).

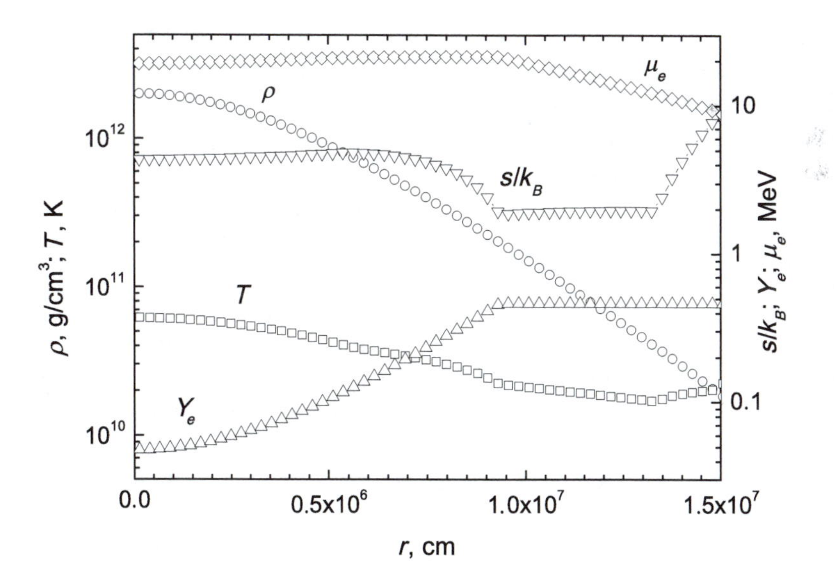

Figure 11.7 Initial model in the equatorial plane in the 2D computations. Shown are the radial dependence of the density ρ, temperature T, entropy per nucleon s/kB, electron number Y_e, and chemical potential of the electrons μ_e [20].

In the collapse problem [14], we obtained two supposedly unstable regions for convection (if we use the Schwarzschild stability criterion excluding neutrinos). The first region near the center of the star exists for $\gtrsim 10$ ms. This region corresponds to the decreasing entropy profile in Fig. 11.7. The second narrow region forms near the accreting shock wave and exists for a long time. Another criterion for the onset of convection for gas dynamics with chemical transformation (neutronization of matter occurs in gas-dynamic time) is the Ledoux stability criterion [71,209]. The Ledoux criterion does not consider neutrino transport, which turns out to be important.

In the 2D computations in spherical coordinates $(r, 0 < \theta < \pi/2)$, we used a 60×30 grid in the computational domain restricted to a region $r \leq 0.6r_{\mathrm{eq}}$, assuming axial symmetry with the plane of symmetry $\theta = \pi/2$. For the neutrino spectrum, we are using a logarithmic grid with 15 intervals up to 40 MeV. Upon approaching the origin of coordinates, the spatial grid is refined in the angular variable, and in the calculations according to an explicit scheme, the Courant constraint is applied to the time step $\Delta t \lesssim \Delta r \Delta\theta/(c_{\mathrm{s}} + |v|)$. The authors of [281] used a less detailed grid in the angles when approaching the origin and combined the intervals by angles. In our calculations, we did not change the grid, but split the time steps of integration by angles upon approaching the origin, i.e., roughened the description in the angular direction near the origin.

Boundary conditions of the problem are as follows. At the outer boundary, there is a non-flow of matter and a smooth wall; neutrinos freely leave the computational domain. In this calculation, we do not aim to describe the ejection of the envelope; if we had set such a goal, we would have chosen a sufficiently large computational domain bounded by a smooth wall for the matter in order to control the calculation results. The mass is conserved with machine accuracy; the total energy losses associated with the inaccuracy of the hydrodynamic description under gravity on a finite grid are much smaller than the neutrino energy losses; the non-conservation of the z-component of the angular momentum is also associated with approximation errors (there are scheme viscosity and friction of layers of the matter) and does not exceed fractions of a percent in the method even for multiple revolutions of the matter in the case of strong rotation [17]. In this calculation, the rotation is weak; the angular velocity in the forming neutron star, due to the loss of neutrino energy, reaches 0.4 rad ms^{-1}, which accounts for less than one revolution in the central part over the entire calculation time.

The computational complexity of the problem lies in the low gas-dynamic velocities of matter and sound in comparison with the light speed of neutrino transport only in the transparent region, $\Delta t \lesssim \Delta r/(c_s + |v|)$. Although diffusion and energy exchange with neutrinos are calculated according to an implicit scheme, there is no possibility to calculate the neutrino transport in one time step through several spatial intervals of the computational grid without loss of accuracy. The opaque region does not limit the time steps of the explicit gas-dynamic scheme. In practice, the time step of gas dynamics is selected based on the requirement of a limited number of time steps of an implicit scheme describing the transport of neutrino energy per one gas-dynamic step. This requirement automatically leads to the time step condition in the transparent region, in which energy is carried by light speed neutrinos and not by all matter (its gas-dynamic speed and the speed of sound are much smaller than the speed of light). Another feature of the problem is the use of a conservative scheme in the presence of a large degeneracy of electrons. In each volume of the spatial grid, to construct a conservative scheme, we use the volumetric energy densities of matter and the spectral density of neutrinos. This solves the problem of the opaque region, in which the time steps are determined by the gas-dynamic transport and the speed of sound, and there is no need to calculate the multiple emission and absorption of neutrinos. However, while the error of determining the energy of matter is small, the resulting temperature error is large, and we can reach a physically unallowable region of parameters of nuclear statistical equilibrium $(\rho, \rho\epsilon, Y_e)$. In practice, on the computational grid used, it is possible to calculate the development of instability for strong degeneracy $\mu_e \sim 60$ MeV at temperature $kT \lesssim 10$ MeV. Finally, gas dynamics include the transport of the number of electrons per nucleon Y_e together with matter: It is a natural independent variable for the Lagrangian description of the problem on a grid moving with matter [14]. It turned out that the high resolution of the Euler scheme on a fixed grid does not always provide physically acceptable parameters in the transition region of neutronization between iron and free nucleons for the equation of state of nuclear statistical equilibrium if the number of electrons per nucleon Δn_e is used as an independent variable. The use

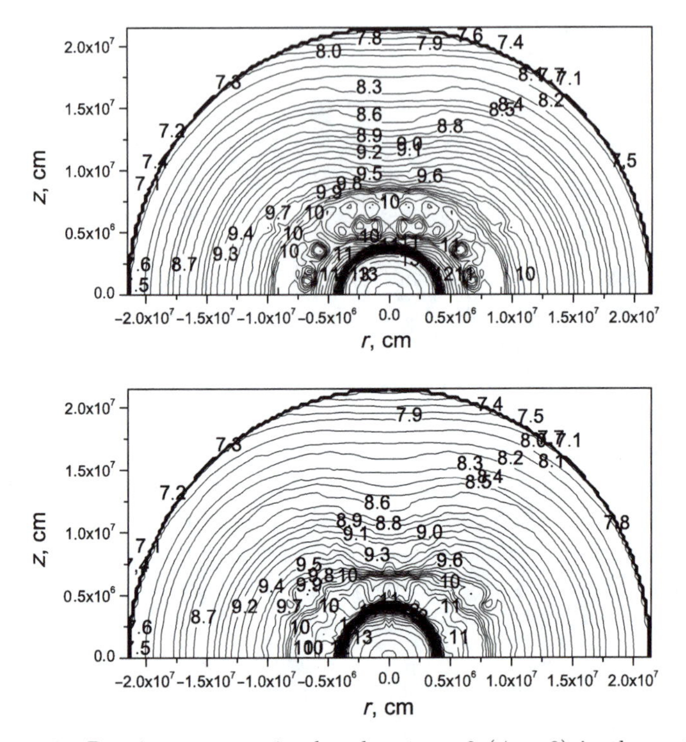

Figure 11.8 Density contours in the plane $y = 0$ ($\phi = 0$) in the problem of collapse of mass 1.4 $M_\odot$ taking account of the transport of neutrinos and their spectrum at times $t = 11.72$ ms (top), $t = 12.59$ ms (bottom) for parameters $\lg \rho_{\min} = 7$, $\lg \rho_{\max} = 13$, $\Delta \lg \rho = 0.1$ (unit of density g cm^{-3}) demonstrates the development of large-scale convection [21].

of the difference in the concentrations of electrons and positrons as the independent variable in Eq. (11.15) eliminates this problem.

The density level lines in Fig. 11.8 at time $t = 11.72$ ms show the development of large-scale convection over the gas-dynamic time $(G\bar{\rho})^{-1/2} = 4$ ms for medium density $\bar{\rho} \approx 10^{11}$ g cm^{-3}. In this case, there is a rearrangement of the outward-decreasing unstable entropy profile. In this part, neutrinos are lost, and convection conditions close to the Schwarzschild condition are realized due to *irreversible neutronization*. The calculations with the rotation taken into account show that perturbations of longer wavelengths are formed contrary to the case without rotation [18,99]. In a more realistic 3D case, rotation helps to form the same long-wavelength perturbations [20,95].

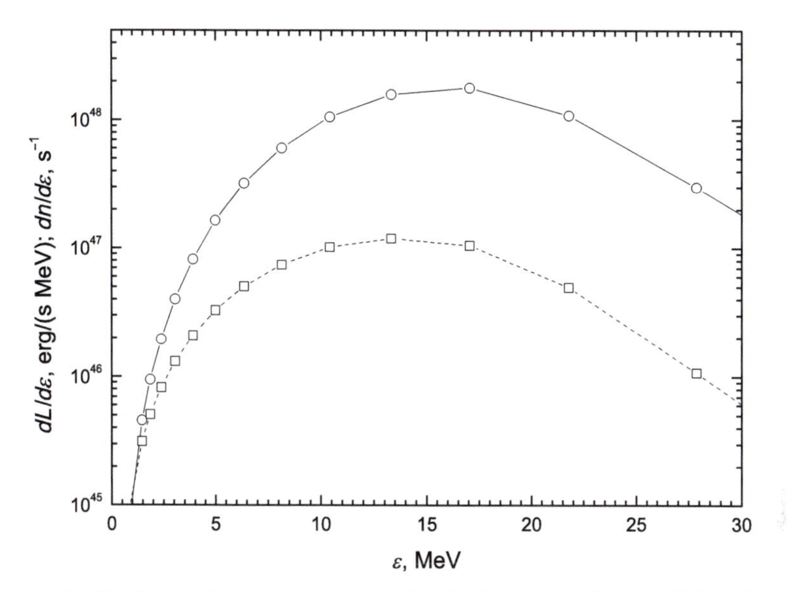

Figure 11.9 Spectral neutrino luminosity during the collapse, $dL/d\epsilon$ (solid), and spectral particle flux, $dn/d\epsilon$ (dashed), at time moment 15.77 ms in the 2D model [20].

In the central part with a high density, neutrinos are captured by matter, neutronization is reversible, and convection develops worse. With neutrino transport turned off, the problem becomes a mathematically rigorous stability problem, and convection in the entire computational domain develops over a long time (100 ms), starting from the peripheral rarefied area. Recalculation of specific entropy under the assumption of fast relaxation of the number of electrons Y_e to the equilibrium value Y_e^{eq} removes the unstable entropy profile.

Figure 11.9 shows the spectral neutrino luminosity during the collapse, $dL/d\epsilon$, and the spectral particle flux, $dn/d\epsilon$, at time $t =$ 15.8 ms for the 2D computations. A spherically symmetric 1D computation yields the mean energy of 10 MeV [14,19]. In the 2D computations, the escaping neutrino spectrum becomes harder due to convection in the central region with high-energy neutrinos that are trapped by optically dense matter in the 1D computations. The energy of the emergent neutrinos is approximately equal to the maximum electron chemical potential of $\approx$ 20 MeV in Fig. 11.7. The mean energy is equal to the ratio of the luminosity and particle flux, $\langle \epsilon_{1\text{D}} \rangle = 15$ MeV. The growth in the energy of the emergent neutrinos

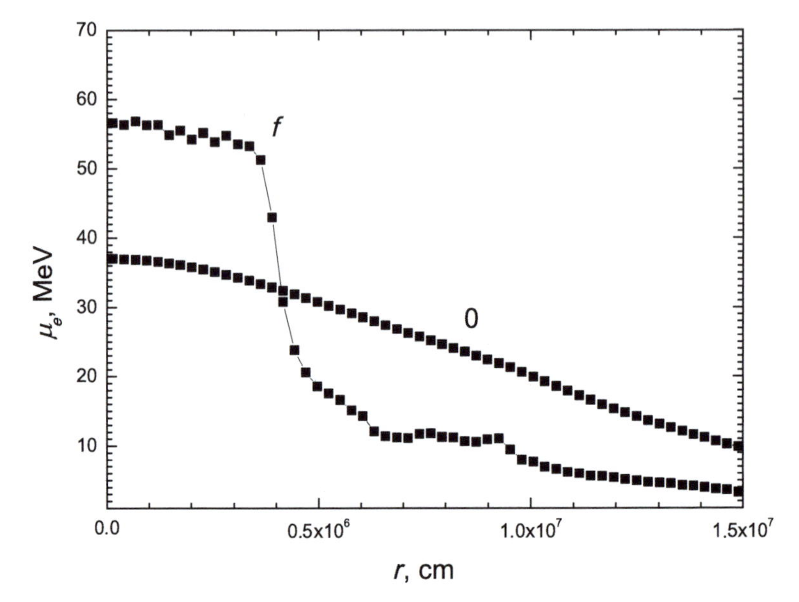

Figure 11.10 Chemical potential of electrons at the beginning of calculations ($t \approx 0$ and in the end of 2D calculations ($t_f = 11.72$ ms) in the equatorial plane [21].

we have obtained is in agreement with old simulations [87]. The maximum spectral luminosity is reached at 18 MeV. As shown by the profile of the chemical potential of electrons (and the chemical potential of neutrinos close to it in the opaque region) in Fig. 11.10, the energy of the outgoing particles approximately corresponds to the maximum values of the chemical potential in the convective region, which does not affect the center with a high chemical potential of ≈ 60 MeV.

In addition, we considered the same task in the 3D case with the neutrino transport without the consideration of spectrum on the computational grid ($60 \times 30 \times 20$) in the region $r < 1.5 \times 10^7$ cm, $0 < \theta < \pi/2$, $0 < \phi < 2\pi$. The density contours for time $t = 16.0$ ms shown in Fig. 11.11 (the results in 3D are close to the axial symmetric 2D case) illustrate the development of large-scale convection over the gas-dynamical timescale $(G\bar{\rho})^{-1/2} = 4$ ms for the mean density $\rho \approx 10^{11}$ g/cm^3. This corresponds to the initial unstable region $ds/dr < 0$ (Fig. 11.7). Neutrinos are lost here, and a convection condition close to the *Schwarzschild condition* arises due to *non-equilibrium neutronization*. Computations taking account

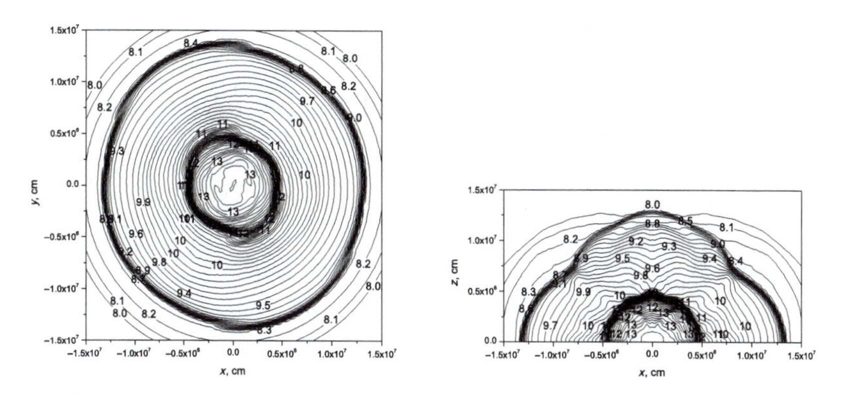

Figure 11.11 Density contours in the $y = 0$ and $x = 0$ planes for the collapse of a mass 1.4 $M_{\odot}$ with the neutrino transfer without taking account of spectrum in 3D calculations at time $t = 15.99$ ms ($\lg \rho_{\min} = 8$, $\lg \rho_{\max} = 13$, $\Delta \lg \rho = 0.1$), demonstrating the development of large-scale convection. 3D calculations demonstrate the conservation of the axial symmetry due to a modest rotation [20].

of rotation demonstrate the development of longer-wavelength perturbations than in the case without rotation [18,99]. Figure 11.11 demonstrates with 3D calculations the formation of the instability in coordinates (r, θ) and the conservation of the axial symmetry along the coordinate ϕ due to a modest initial rotation. This effect of the rotation can be important to explain the contradictory results of other authors. Results [111,318] demonstrate the large-scale character of convection and lead to the SN explosion, while improved 3D computations cast doubt on large-scale convection and the explosion [112,263].

We have considered the development of large-scale convection in a collapsing stellar core with a modest rotation taking account of the spectral neutrino transport. Convection arises in the central region over the gas-dynamical timescale, ~10 ms, due to non-equilibrium neutronization, with the loss of some fraction of the neutrinos. The mean energy of the escaping neutrino is 15 MeV. There is no doubt that observations of neutrinos from SNe are important for confirming models for the collapse and explosion. Only 20 neutrino events were registered for close SN 1987A. The first set of publications about this event indicate high energies: 20–40 MeV IMB [64], 9–35 MeV Kamiokande-II [161], and 20 MeV Baksan-LSD [32], closer to the large-scale convection model than to expectations for a spherically symmetric collapse.

Comparing the results of this study with modern gas-dynamic models (without the magnetic field) of foreign authors, we should note the use of the simplified neutrino transport in 3D models instead of solving Boltzmann's equations. These authors use a more accurate equation of state at nuclear densities, a full set of reactions of neutrino–matter interaction and sometimes include the effects of general relativity, which are not important when the size of the forming proto-neutron star is tens of kilometers. The main emphasis of these models is placed on explaining the envelope ejection without a detailed explanation of the physics of the processes. In particular, the key question about the scale of convection in the accretion region and conclusions about the ejection of the SN envelope turn out to depend on the adopted dimension of the mathematical problem. In 2D models [121,232], the SN explosion occurs simultaneously with large-scale convection. In a more realistic 3D setting [112,263], convection in the accretion region remains small scale, and the required energy input into the presupernova envelope often does not occur [86]. In our analysis, this means an insufficient effect of small-scale convection in the 3D setting, primarily on the energy of outgoing neutrinos. Our calculations of collapse in 3D taking account of neutrino transport clearly show that a small initial rotation turns the 3D problem into a 2D setting and eliminates the problem of the development of large-scale perturbations [20–22].

The axially symmetric calculations of the authors of [121,238] demonstrate the average neutrino energies of 12–20 MeV and the neutrino luminosity tens of percent higher than in the spherically symmetric model. Recall that in a spherically symmetric model, an artificial increase in the average neutrino energy to 30 MeV would be sufficient to explain an SN even with a constant neutrino luminosity [16,52]. Most 3D models [112,263] are focused on the analysis of the development of instability, while neutrino spectra are not given. However, studies have recently appeared on the possibility of detecting neutrinos from collapsing supernovae, for which the calculation of the neutrino spectrum is obligatory. During the explosion of the last close SN 1987A in a neighboring galaxy in 1987, several neutrino events associated with the subsequent supernova explosion were detected (photons arise from the extended pre-supernova shock wave propagating through matter and are observed later than neutrinos from the gravitational collapse of a compact central mass). Modern observatories, for example, Super-Kamiokande, with a large amount of working substance promise a thousand times more detections of neutrino events from a nearby supernova and even the possibility of detecting neutrinos from distant collapsing supernovae at a distance of 100 kpc [3]. The calculated neutrino luminosity (the total energy is comparable to the gravitational energy of a neutron star and is emitted for a few seconds), together with the neutrino spectrum (typical average energies of 10–20 MeV), gives the detector response depending on the distance from the Earth. The detection rate of neutrino events turns out to be distinguishable for the calculated collapse models with different neutrino spectra. However, the uncertainty in the calculation of neutrino oscillations introduces uncertainty in the prediction of the rate of neutrino events. The average neutrino energies at the level of 5–10 MeV give a significantly lower rate of detected events and always turn out to be distinguishable from 15 MeV neutrinos (low-energy neutrinos are harder to detect), and the average energies of 30 MeV neutrinos may turn out to be indistinguishable from 15 MeV neutrinos [3]. We argue that the key to verifying the large-scale convection model is the ability to experimentally distinguish the average neutrino energies of 15 MeV from 30 MeV. A mere detection of neutrino events is not sufficient to verify the theory of explosion; a neutrino spectrum is needed. It turns out that the time-integrated neutrino spectrum, even if neutrino oscillations are allowable, can be reproduced by detecting neutrinos with several types of detectors [237].

11.4 Conclusion

Large-scale convection in the problem of the collapse of a stellar core with weak rotation has been considered, taking into account the transfer of neutrinos and their spectrum. Convection occurs in the central part over the gas-dynamic time of milliseconds due to non-equilibrium neutronization when neutrinos escape this region. The isolation of the long-wavelength mode is facilitated by the small initial rotation, since the initial configuration differs from the spherical one. The obtained value of the average neutrino energy of

15 MeV is 1.5 times higher than the value of spherically symmetric calculations and will increase the energy transferred from the neutrino to the expanding envelope by a corresponding number of times. This effect is essential for explaining the SN explosion.

The conditions for the stability of the Schwarzschild gas dynamics are realized near the forming neutrinosphere in the region of nonequilibrium neutronization. In the very center with trapped neutrinos with high chemical potential MeV, neutronization is reversible and convection is suppressed in accordance with the Ledoux criterion.

Considering the processes of interaction of neutrinos with the envelope in our model is complicated by a significant increase in the CPU time in proportion to the cube of the number of energy intervals of computational grid. The implicit scheme describing the reactions of weak interaction calculates inverse matrix of Jacobians of the ordinary differential equations system. Probably, both an increase in average energy and an increase in luminosity due to large-scale convection are important to explain the ejection of the supernova envelope with the necessary energy deposition of 10^{51} erg.

Additionally, we proved the acceptance of the axially symmetric consideration for the rotating core by the direct 3D hydrodynamic simulations with the neutrino transport without the consideration of neutrinos spectrum.

Chapter 12

GR Effects in SN and Binary Systems

The *coalescence of components of a binary star* with equal masses ($M_1 = M_2 = M_\odot$) and moving in circular orbits is considered [15]. The equation of state for degenerate neutrons is used, leading to the equation of state for an ideal gas. The initial model has zero temperature, corresponding to a polytrope with $n = 1.5$. To reduce the required computational time, the initial close binary is constructed using the self-consistent field method. The computations use Newtonian gas dynamics, but the back reaction of the gravitational radiation is taken into account in a PN2.5 post-Newton approximation, obtained using ADM formalism. This makes it possible to apply previous experience of constructing high-order Godunov-type difference schemes, which are suitable for end-to-end calculations of discontinuous solutions of the gas-dynamics equations on a fixed Eulerian grid. The Poisson equations were solved using an original spherical-function expansion method. The 3D computations yielded the parameters of the gravitational signal. Near the radiation maximum, the strain amplitude is $rh \sim 4 \cdot 10^4$ cm, the power maximum is $4 \cdot 10^{54}$ erg s^{-1}, and the typical radiation frequency is $\gtrsim$1 kHz. The energy carried away by gravitational waves is $\gtrsim 10^{52}$ erg. These parameters are of interest, since they form an inherent part of a rotational mechanism for the supernova explosion. They are also of interest for the planning of gravitational-wave detection experiments.

12.1 Introduction

Close compact binaries are observed in nature (the binary neutron stars PSR B1913+16, PSR J0737-3039, and PSR J0437-4715 and the binary white dwarf RX J0806). At least one-fifth of the stars are components of close binaries. The best known binary pulsar PSR B1913+16 [163] is located at a distance of 7 kpc. It is expected that there are about 1000 such binaries in the galaxy. High-precision measurements agree with the expectations of *general relativity (GR)* [201], which predicts the radiation of gravitational waves, decrease in the orbital separation, and a finite lifetime for such systems (e.g., the decrease of the orbital period of PSR B1913+16 by 76 mks per year). D. Taylor and R. Hulse received Nobel Prize in 1993 for observations of PSR B1913+16. Mergers of neutron stars are accompanied by perturbations of the gravitational-field metric, i.e., by gravitational waves propagating at the speed of light. It has also been suggested that mergers of neutron stars are sources of short gamma-ray bursts [1].

Gravitational waves have a quadrupole nature, namely, they are generated by the third time derivative of the reduced quadrupole moment of the mass of the system with the trace equal to zero [149,227,279]. Thus, gravitational waves are not emitted by spherically symmetric bodies or stationary axially symmetric bodies. However, they may be emitted by non-stationary, axially symmetric bodies. The strongest sources are objects with large masses and accelerations, namely, close compact stars (neutron stars or black holes in binaries), colliding galaxies, and the Universe at the epoch of the Big Bang (relict gravitational waves).

Various gravitational wave detectors, or gravitational telescopes, exist [227]. A gravitational wave changes the distance between test masses. There are two common types of detectors. One is mechanical detector, *Weber gravitational antenna*. These are massive metallic rods cooled to low temperatures. Variations in the length of the rods by 10^{-16} m are registered by piezotransducers. The resonance of its own oscillations and the frequencies of the gravitational waves are important for the action of such detectors. Another type of detector is based on the use of lasers to measure the variation of the distance between test masses with a Michelson interferometer. This type of detector was suggested in Ref. [139]. The best known currently active detector is the *Laser Interferometer Gravitational-Wave Observatory* (LIGO) [2,49,277], which has two detectors with 4 km long L-498-shaped arms. Its working range is 10–10^4 Hz, and the target sensitivity to the strain amplitude is 10^{-21} or, in terms of the distance variations, 10^{-21} m. The sensitivity of the detector suffers from seismic and thermal noise. In addition to the LIGO-project, the VIRGO detector, which has 3 km arms, an effective detector of length 120 km, and an accuracy of 10^{-22}, is also operational. The LISA space mission planned for launch after 2020 will have mirrors placed in circumsolar orbits with 5 million kilometer arms. It is designed to be sensitive to displacements of 20 pm in 10^{-4}–10^{-1} Hz [48]. The first gravitational event has been detected in September 2015. Supernova SN 1987A in a nearby galaxy enriched astronomical measurements by providing the first detection of neutrinos from a supernova [175]. Modern gravitational wave detectors should be sensitive to such nearby by supernovae. The sphericity of the explosion of SN 1987A is indicated by the absence of a compact remnant, as follows from estimates of the accretion rate and comparison with the observed photon radiation [54].

We and our colleagues are studying multidimensional effects in the gravitational collapses and Types I and II supernovae. In particular, we have discussed and justified a fundamentally non-1D mechanism for a supernova explosion based on magneto-rotational and rotational models [60,65,168,266]. The idea behind this approach is the amplification of rotation-related effects or frozen-in magnetic field as the dimensions of the system decrease during the collapse. In particular, this work suggests that a rotational explosion should include (i) the formation of a rapidly rotating $2\,M_\odot$ neutron star during the collapse, (ii) the separation of this neutron star into two components, and (iii) the evolution of the resulting binary neutron star, with energy losses via gravitational-wave radiation and the subsequent coalescence of the neutron stars or mass-transfer onto the more massive component. We do not wish to dwell on the fundamental difficulties encountered by both scenarios, such as the problem of the separation of the neutron star in the rotational mechanism or the low kinetic energy released by the explosion of a low-mass neutron star [110]. All these models require multidimensional gas-dynamical or magneto-dynamical computations with good resolution. In our opinion, a more common direction for studies of supernovae is the investigation of large-scale convective motions during the explosion [52,58,88,101]. This model is supported by the growth of the neutrino cross-section for scattering on electrons in the supernova envelope with increasing neutrino energy, found in detailed spherically symmetric computations of the collapse [14].

A key issue in the rotational mechanism for the explosion is the formation of a binary neutron star [176] and its evolution. This is the object of our study. The lifetime of the system is determined by two parameters: the masses of the two components. The initial value of the momentum becomes unimportant, since momentum is carried away by gravitational radiation. The most interesting moment is when the components approach each other at the peak of the gravitational radiation. The equation of state determines the mass distribution in the close binary and influences the gravitational signal, as well as the evolution of the binary [197]. Analyzing this problem includes constructing an initial stationary model for a close binary, choosing a mathematical model that incorporates the gas-dynamical equations, including the gravitational radiation back reaction in these equations, and developing a numerical method for the integration of the three-dimensional (3D) equations. The main results are the data on the expected gravitational radiation. This is important information for planning experiments aimed at detecting gravitational waves and identifying the explosion mechanism. We plan to solve multidimensional physical problems related to the collapse and the development of instability at the boundary of the hot region opaque to neutrinos by applying the methods of multidimensional gas dynamics we have developed, including the effects of neutrino transfer (diffusion with flux limiters) [52].

The simplest way to take into account gravitational radiation in slowly moving sources is applying a correction to the potential that is proportional to the fifth derivative of the reduced quadrupole moment of the mass [227,279]. In this case, the delay is not taken into account, but the problem arises of numerically computing the derivative. Solutions of the merger problem in the framework of general relativity are beginning to appear [128,212]. The computational complexity of the problem for the stars stems simultaneously from the 3D formulation of the problem, the large density gradients, and the need to carry out the computations over the prolonged evolution of the system up to the merger (many periods of the binary system, without loss of momentum due to the viscosity of the scheme). Thus, it is necessary to have a computational grid with high resolution (enabling resolution of the contact discontinuity over a small number of intervals) and small numerical diffusion. The results of the computations and the dependence of the shape of the gravitational signal in the vicinity of the maximum on the rigidity of the equation of state are quite sensitive to the numerical diffusion, as was first studied in Ref. [197]. Another fundamental problem for the numerical solution is the formulation of the general-relativity equations as a Cauchy problem. The formalism of Arnowitt, Deser, and Misner (ADM) [39–47] is convenient for this. The equations of motion are rewritten in Hamiltonian form and the equations of motion for the metric and conjugate momentum obtained. In the present stage of our studies, we have chosen a simpler approach, using the PN2.5 post-Newtonian expansion [72], which is obtained using the *ADM-formalism.* We then applied Newtonian hydrodynamics, taking into account the gravitational-radiation back reaction in the PN2.5 post-Newtonian approximation [270–272]. We applied the PPM scheme to solve the gas-dynamical equations [108]. Allowing for the gravitational-radiation back reaction reduces to the solving additional Poisson equations using our own original method [10] and Chapter 9 of Part 2.

12.2 Formulation of a problem

We used the equations from Ref. [72]. The conservation of baryon mass (henceforth, mass) can be expressed as

$$\frac{\partial \rho}{\partial t} + \frac{\partial \rho v^j}{\partial x^j} = 0, \tag{12.1}$$

the equation of motion is

$$\frac{\partial w^i}{\partial t} + v^j \frac{\partial w^i}{\partial x^j} = -\frac{1}{\rho}\frac{\partial P}{\partial x^i} - \frac{\partial \psi}{\partial x^i} - \frac{\partial \psi_r}{\partial x^i}, \tag{12.2}$$

and the equation for the specific internal energy is

$$\rho\frac{d\epsilon}{dt} + P\frac{\partial v^i}{\partial x^i} = 0, \quad \frac{d}{dt} \equiv \frac{\partial}{\partial t} + v^i\frac{\partial}{\partial x^i}. \tag{12.3}$$

The potential ψ is derived from the Poisson equation

$$\Delta\psi = 4\pi G\rho. \tag{12.4}$$

The potential responsible for the gravitational-radiation back reaction is

$$\psi_r = \frac{2}{5}\frac{G}{c^5}\left(R - \dddot{\mathcal{Q}}^i{}_j x^j \frac{\partial\psi}{\partial x^i}\right). \tag{12.5}$$

The 3D velocity of the matter v_i (dynamical velocity) and specific momentum w_i (kinematic velocity) are related by the expressions

$$v^i = w^i + \frac{4}{5}\frac{G}{c^5}\dddot{\mathcal{Q}}^i{}_j w^j. \tag{12.6}$$

We used the symmetric tensor with zero trace

$$\dddot{\mathcal{Q}}_{xy} = \mathrm{STF}\left[\dddot{Q}_{xy}\right] = \frac{1}{2}\dddot{Q}_{xy} + \frac{1}{2}\dddot{Q}_{yx} - \frac{1}{3}\delta_{xy}\dddot{Q}_{zz}. \tag{12.7}$$

In curvilinear coordinates,

$$\begin{aligned}\dddot{Q}^i{}_j &= 2\int d\mathbf{r}\rho\left(-w^i\left(3\partial_j\psi + 2\frac{\partial_j P}{\rho}\right) - x^i(w^s\partial_{sj}\psi + \partial_j\tilde{\psi})\right)\\ &= 2\int d\mathbf{r}\Big(-2(\mathrm{grad}P)_j w^i + (\mathrm{grad}\psi)_j(x^i\mathrm{div}\rho\mathbf{w} - 2\rho w^i)\\ &\qquad - \rho x^i(\mathrm{grad}\tilde{\psi})_j\Big),\end{aligned} \tag{12.8}$$

these are third derivatives of the quadrupole moment with accuracy to within $(v/c)^2$. The remaining potentials are derived from the additional Poisson equations

$$\Delta\tilde{\psi} = -4\pi G\mathrm{div}(\rho\mathbf{w}), \tag{12.9}$$

$$\Delta R = 4\pi G\dddot{\mathcal{Q}}^i{}_j x^j(\mathrm{grad}\rho)_i. \tag{12.10}$$

Instead of the motion equation for the velocity (12.2), it is more convenient to apply momentum conservation:

$$\frac{\partial \rho w^i}{\partial t} + \frac{\partial(\rho w^i v^j + P\delta^{ij})}{\partial x^j} = -\rho \frac{\partial(\psi + \psi_r)}{\partial x^i}, \tag{12.11}$$

and instead of the equation for the specific internal energy (12.3), to apply the equation for the energy density $E = \epsilon + \frac{1}{2} w^i w^i$

$$\frac{\partial \rho E}{\partial t} + \frac{\partial(\rho E + P)v^j}{\partial x^j} = -\rho v^i \frac{\partial(\psi + \psi_r)}{\partial x^i} + \frac{4}{5}\frac{G}{c^5} \dddot{Q}^i{}_j w^j \times \left(\rho \frac{\partial(\psi + \psi_r)}{\partial x^i} + \frac{\partial P}{\partial x^i}\right). \tag{12.12}$$

When gravitation and the gravitational-radiation back reaction are taken into account, mass and momentum are conserved [72], i.e.,

$$\frac{d}{dt} \int d\mathbf{r} \rho = 0 \tag{12.13}$$

and

$$\frac{d}{dt} \int d\mathbf{r} \rho w^i = 0, \tag{12.14}$$

while the gravitational radiation carries away momentum and energy, i.e.,

$$\frac{d}{dt} \int d\mathbf{r}(x^i \rho w_j - x^j \rho w_i) = \frac{2}{5}\frac{G}{c^5} \left(\dddot{Q}^i{}_s \ddot{Q}^s{}_j - \dddot{Q}^j{}_s \ddot{Q}^s{}_i\right) \tag{12.15}$$

(the summation on the right-hand side is over s only) and

$$\frac{d}{dt} \int d\mathbf{r} \left(\rho E + \frac{1}{2}\rho\psi\right) = -\frac{1}{5}\frac{G}{c^5} \left(\dddot{Q}^i{}_j \dddot{Q}^j{}_i\right). \tag{12.16}$$

The independent variables of the problem are ρ, ρw^i, ρE. We worked in spherical coordinates, which are convenient for solving the Poisson equation. In the orthogonal curvilinear coordinates x^i, we have $\Pi_{ij} = \rho v_i w_j + P\delta_{ij}$, and the gradient of the scalar ψ, the divergence of the vector $\mathbf{a}$, and the divergence of the tensor Π can be written (all vectors are normalized in the usual way, such that all basic vectors have unit length, see, e.g., Ref. [264]:

$$(\mathrm{grad}\psi)_i = \frac{1}{\sqrt{g_{ii}}}\frac{\partial \psi}{\partial x^i},$$

$$\mathrm{div}\mathbf{a} = \frac{1}{\sqrt{|g|}}\sum_i \frac{\partial}{\partial x^i}\left(\frac{\sqrt{|g|}a_i}{\sqrt{g_{ii}}}\right),$$

$$(\mathrm{Div}\Pi)_j = \frac{1}{\sqrt{g_{jj}}}\sum_i \left(\frac{1}{\sqrt{|g|}}\frac{\partial}{\partial x^i}\left(\frac{\sqrt{|g|g_{jj}}\Pi_{ij}}{\sqrt{g_{ii}}}\right) - \Pi_{ii}\frac{\partial \ln \sqrt{g_{ii}}}{\partial x^j}\right),$$

where $|g| \equiv \det\, g_{ij}$, $g_{ij} \equiv \left(\frac{\partial \mathbf{r}}{\partial x^i}, \frac{\partial \mathbf{r}}{\partial x^j}\right) = \mathrm{diag}(g_{11}, g_{22}, g_{33})$ is the metric tensor, and the vector $\mathbf{r}$ is written in rectangular Cartesian coordinates, $\mathbf{r} = r(\sin\theta\cos\phi, \sin\theta\sin\phi, \cos\theta)$. The tensor $\dddot{Q}_{xy}$ can be considered in Cartesian coordinates, and the summation with this tensor can also be performed in rectangular coordinates. It is possible to use vectors normalized to unit vectors in this tensor and to avoid transforming to Cartesian coordinates to obtain a symmetric tensor with zero trace. The coordinates of the vectors are related as

$$w_x = t_{xi}w_i,$$

where t^x_i is the transformation matrix, equal to

$$T = \begin{pmatrix} \sin\theta\cos\phi & \cos\theta\cos\phi & -\sin\phi \\ \sin\theta\sin\phi & \cos\theta\sin\phi & \cos\phi \\ \cos\theta & -\sin\theta & 0 \end{pmatrix}.$$

The orthogonal matrix is $(t^{-1})_{ix} = t_{xi}$. In spherical coordinates, the vector is $x_i = (t^{-1})_{ix}x_x = (t^{-1})_{ix}(r\sin\theta\cos\phi, r\sin\theta\sin\phi, r\cos\theta)^T = (r, 0, 0)^T$. The symmetry of the tensor does not depend on the coordinate system: $w_{yx} = t_{yi}t_{xj}w_{ij} = t_{xj}t_{yi}w_{ji} = w_{xy}$, the trace of the vector is also an invariant: $w_{xx} = t_{xi}t_{xj}w_{ij} = t_{xi}(t^{-1})_{jx}w_{ij} = \delta_{ij}w_{ij} = w_{ii}$.

To solve this problem, a *splitting over physical processes and directions* is carried out. We use a fixed grid. The radiation transfer and gravitational-radiation back reaction were considered separately. We integrated the gas-dynamical equations using the PPM method [108], a second-order Godunov scheme. The Poisson equation was solved in spherical coordinates using the method suggested in Ref. [10]. Therefore, the gas-dynamical equations were also solved in spherical coordinates. In turn, we applied the method of variable directions when solving the gas-dynamical equations: The integration was made in several steps, with the derivatives conserved along one direction only at each step. Since the initial equations for the algorithm are in divergence form, the scheme enables accurate conservation of the mass and momentum. Energy is conserved in the absence of gravity only. The scheme is explicit and has a time step limited by the Courant condition, which is important for integration over angles at small r. It is possible to circumvent this condition by making several integration steps over angles for each step of the integration along r when r is small.

12.2.1 *Equation of state and initial model*

We used the equation of state for a Fermi gas of non-interacting neutrons [202]:

$$n = 2\left(\frac{1}{2\pi\hbar}\right)^3 \int_0^\infty \frac{4\pi p^2 dp}{1 + \exp\frac{c\sqrt{(m_n c)^2 + p^2} - \mu_n}{kT}}, \tag{12.17}$$

$$\rho\epsilon = 2\left(\frac{m_n c}{2\pi\hbar}\right)^3 \int_0^\infty \frac{4\pi(c\sqrt{(m_n c)^2 + p^2}))p^2 dp}{1 + \exp\frac{c\sqrt{(m_n c)^2 + p^2} - \mu_n}{kT}}, \tag{12.18}$$

$$P = c^2\left(\frac{m_n c}{2\pi\hbar}\right)^3 \int_0^\infty \frac{4\pi\frac{p^2}{3(c\sqrt{(m_n c)^2 + p^2})}p^2 dp}{1 + \exp\frac{c\sqrt{(m_n c)^2 + p^2} - \mu_n}{kT}}. \tag{12.19}$$

The neutrons were taken to be non-relativistic: $p \ll m_n c$. Therefore, for any temperature, the pressure and density are related as in an ideal gas:

$$P = \left(\frac{5}{3} - 1\right)\rho\epsilon, \tag{12.20}$$

while we have a polytropic equation in the initial state at zero temperature:

$$P = K\rho^{1+1/n}, \quad K = \frac{3^{2/3}\pi^{4/3}\hbar^2}{5m_n^{8/3}}\rho^{5/3}, \quad n = 1.5. \tag{12.21}$$

Of course, we understand that a real neutron star contains neutrons, protons, and electrons and it is necessary to take into account the interaction of the baryons, as was done, e.g., in Refs. [154,192]. Effectively, this can be done by changing the constants K and Γ [197]. The influence of these parameters on the characteristics of gravitational radiation was also studied for the first time in Ref. [197].

We applied the *self-consistent field method* [12,150,250] to compute the initial model, a close neutron star binary. The momentum equations without meridional circulation, $\mathbf{v} = (0, \Omega\varpi, 0)$, in cylindrical coordinates (φ, ϕ, z) are as follows:

$$\frac{\partial P}{\partial \varpi} - \frac{\rho v_\phi^2}{\varpi} = -\rho\frac{\partial \psi}{\partial \varpi}, \tag{12.22}$$

$$\frac{1}{\varpi}\frac{\partial P}{\partial \phi} = -\frac{\rho}{\varpi}\rho\frac{\partial \psi}{\partial \phi}. \tag{12.23}$$

These equations show that $\frac{\partial v_\phi}{\partial z} = 0$, i.e., $v_\phi = v_\phi(\varpi)$. Introducing the enthalpy $H = \int^{P(\rho)} \frac{dP}{\rho} = K(1+n)\rho^{1/n}$ and centrifugal potential $\Psi = -\int \Omega^2(\varpi)\varpi d\varpi$, we obtain Bernoulli's equation instead of the differential equations of motion:

$$H + \psi + \Psi = C_i, \tag{12.24}$$

where $C_i = \text{const}$ is a constant, that is specific to both the components of the binary. The angular velocities of the two components are

$$\Psi = -\Omega_0^2\frac{\varpi^2}{2}. \tag{12.25}$$

Let us construct a system of two equal-mass stars, $M_1 = M_2 = M_\odot$. It is obvious that $C_1 = C_2$ for equal masses. The Poisson equation for the gravitational potential is valid everywhere:

$$\Delta\psi = 4\pi G\rho. \tag{12.26}$$

It is convenient to apply the efficient method [10], in which an integral representation of the potential is used, to compute the potential from the known density distribution. As in Refs. [150,152], we used the self-consistent field method and ordinary iterations to simultaneous solve the Bernoulli and Poisson equations.

We introduce a grid in spherical coordinates, fixing the point with the maximum distance from rotation axis, i.e., we specify the distance R. We then specify the mass M, the zeroth approximation for the density ρ, the point inside the star closest to the rotation axis A, and the distance R_A. The iteration process appears as follows:

(1) Determine the gravitational potential ψ.
(2) Determine the angular velocity $\Omega_0 = \sqrt{\frac{\psi_R - \psi_A}{(R^2 - R_A^2)/2}}$.
(3) Determine the constant $C = \psi_R + \Psi_R$.
(4) Determine the new density distribution $\rho = H^{-1}(C - \psi - \Psi)$.
(5) Determine the mass: if the fitted mass higher than the previous value, the point A is moved outward by a grid step; if the fitted mass is lower, the distance R_A is reduced by a grid step.

12.3 Results for the coalescence of binary components with a total mass $2M_\odot$

As an example, let us consider a *binary neutron star* with equal mass components $M_1 = M_2 = M_\odot$, initial momentum $J_z = 3.8 \cdot 10^{49}$ erg s, and total energy $E = -1.74 \cdot 10^{53}$ erg. Figure 12.1 shows density contours in the equatorial plane $z = 0$

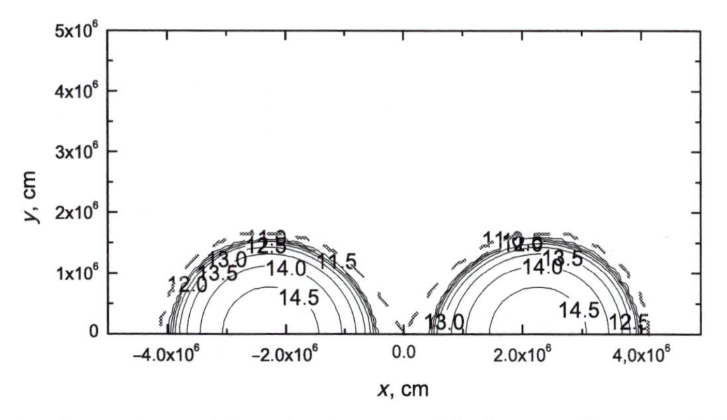

Figure 12.1 Contours of $\log \rho$ in the equatorial plane $z = 0$ at the initial time. The dashed lines delimit the Roche lobe [15].

at the initial time. The two stars form a close system. The dashed lines delimit the *Roche lobes* of the components — the regions of influence of each star. This region is bounded by the equipotential surface of the total potential (gravitational and centrifugal) passing through the first Lagrangian point L_1 (a rotating test mass is in equilibrium at L_1).

The problem of choosing the initial momentum was aimed at resolving the closest system for the $75 \times 75 \times 120$ grid used within the computational domain $0 < r < 5 \cdot 10^6$ cm, $0 < \theta < \pi/2$, $0 < \phi < \pi$.

Further, the components approach, mass transfer occurs, and a new rotating neutron star forms, with some gas expelled to the periphery. This is shown clearly by contour plots in the equatorial plane after several revolutions (Fig. 12.5), before the coalescence of the neutron stars (close to the maximum of the gravitational radiation; Fig. 12.6), and at the end of the computations (Fig. 12.7).

Figure 12.2 shows the time dependence of the strain amplitude components of the metric tensor h_+ and h_x, computed in the quadrupole approximation [9]:

$$h_{ij}^{TT}(\mathbf{n}, t) = \frac{2G}{rc^4} P_{ijkl}(\mathbf{n}) \ddot{\mathfrak{Q}}_{kl}(t - r/c), \tag{12.27}$$

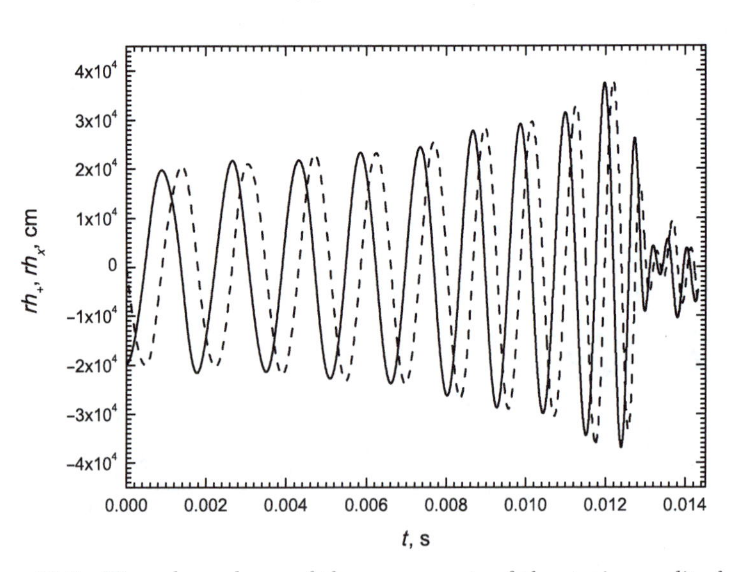

Figure 12.2 Time dependence of the components of the strain amplitude rh_+ (solid) and rh_x (dashed) for an observer located on the symmetry axis [15].

where the operator for projection onto a plane orthogonal to the direction of propagation of the gravitational wave is

$$P_{ijkl}(\mathbf{n}) = (\delta_{ik}-n_i n_k)(\delta_{jk}-n_j n_l)-\frac{1}{2}(\delta_{ij}-n_i n_j)(\delta_{kl}-n_k n_l). \quad (12.28)$$

In the direction of the rotation axis, the non-zero components of the tensor h_{ij}^{TT} are equal to

$$h_{+} = h_{11}^{TT} = -h_{22}^{TT} = \frac{G}{rc^4}\left(\ddot{\mathfrak{Q}}_{11} - \ddot{\mathfrak{Q}}_{22}\right), \quad (12.29)$$

$$h_{x} = h_{12}^{TT} = h_{21}^{TT} = \frac{2G}{rc^4}\ddot{\mathfrak{Q}}_{12}. \quad (12.30)$$

For two point masses M_1 and M_2 with circular orbits of radii a_1 and a_2, the amplitudes of the gravitational waves are

$$rh_{+} = -\frac{G}{c^4}\mu a^2 4\Omega^2 \cos 2\Omega t, \quad (12.31)$$

$$rh_{x} = -\frac{G}{c^4}\mu a^2 4\Omega^2 \sin 2\Omega t, \quad (12.32)$$

where the orbital angular velocity is $\Omega = \sqrt{\frac{G(M_1+M_2)}{a^3}}$, the reduced mass is $\mu = M_1 M_2/(M_1 + M_2)$, and the distance between the point masses is $a = \frac{M_1 a_1 + M_2 a_2}{\mu}$. The gravitational radiation is constant over several revolutions and is well approximated by the point-mass model. When system is then restructured, the gravitational radiation first increases and then strongly falls off. The amplitude is very important for observations, as well as the typical frequency of the radiation ~1 kHz, which is $\Omega = 3.5 \cdot 10^3\ \mathrm{s}^{-1}$ close to the maximum.

The radiation maximum at $t = 0.012$ s is especially clearly visible in Figure 12.3, which also shows the power of the gravitational radiation,

$$\frac{dE_{rad}}{dt} = \frac{1}{5}\frac{G}{c^5}\left\langle \dddot{\mathfrak{Q}}_{ij}\dddot{\mathfrak{Q}}_{ij}\right\rangle, \quad (12.33)$$

and the total emitted energy, which does not change in late phases and reaches $\gtrsim 10^{52}$ erg. The total emitted gravitational energy comprises a significant fraction of the total energy of the system, $\sim 10^{53}$ erg.

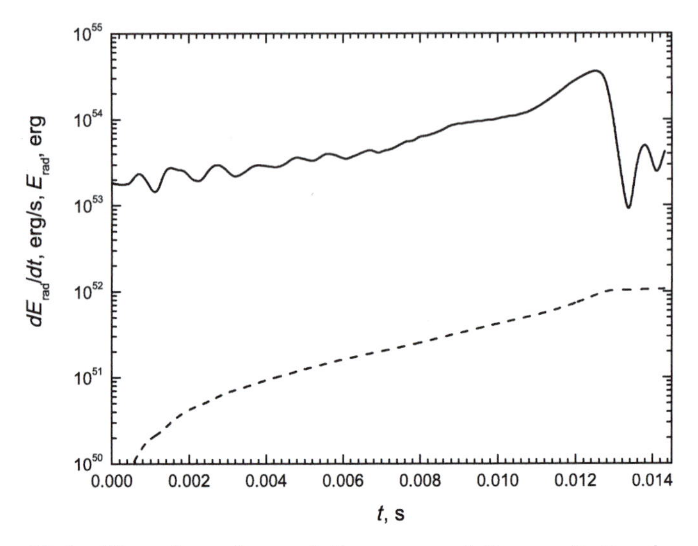

Figure 12.3 Time dependence of the power of the gravitational radiation dE_{rad}/dt (solid) and E_{rad} (dashed) [15].

Figure 12.4 shows the rate of momentum loss,

$$\frac{dJ_{i,rad}}{dt} = \frac{2}{5}\frac{G}{c^5}\epsilon_{ijk}\left\langle \ddot{\mathcal{Q}}_{jm}\dddot{\mathcal{Q}}_{km}\right\rangle, \tag{12.34}$$

and the total lost momentum $\sim 10^{49}$ erg s.

When our numerical scheme is used, the conservation of energy and momentum is computed more precisely than the losses to gravitational radiation. However, the non-conservativeness of the scheme is comparable to the gravitational-radiation losses at $t = 0$. Close to the maximum radiation, the power of the gravitational radiation exceeds the losses due to the non-conservativeness of the scheme with a large margin, i.e., the parameters of the signal close to the maximum should be computed well. The non-conservativeness of the scheme can be observed indirectly by estimating the real number of revolutions of the system prior to coalescence. For our grid, the excess corresponding to the Courant condition was set equal to 0.03. The computations covered 10^5 time steps and required a month of time on a modern scalar computer. The accuracy of the computations could be increased only by parallelizing the computations. This is possible since the scheme has second order in accuracy. Doubling the number of nodes in each direction would increase the accuracy of

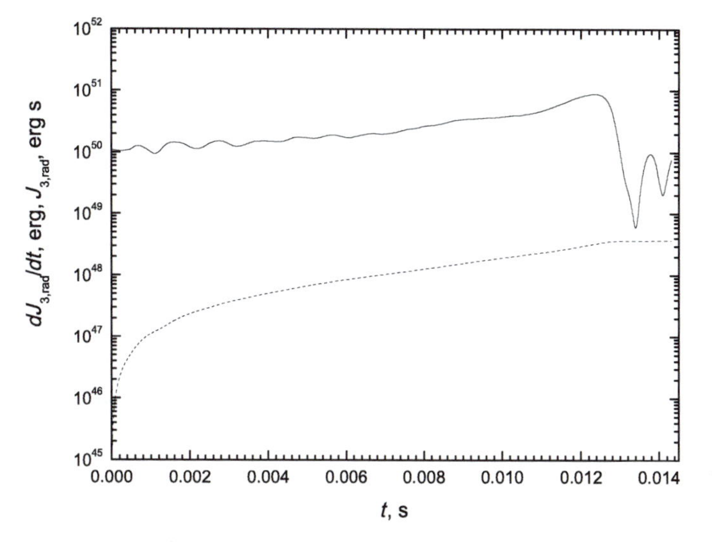

Figure 12.4 Time dependence of the momentum loss rate $dJ_{3,\mathrm{rad}}/dt$ (solid) and lost momentum $J_{3,\mathrm{rad}}$ (dashed) [15].

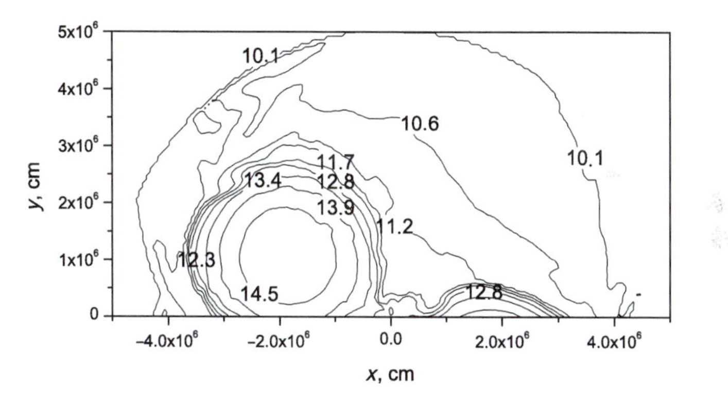

Figure 12.5 Contours of $\lg \rho$ in the equatorial plane $z = 0$ at a time close to the initial time, $t = 3.216 \cdot 10^{-3}$ s [15].

the computations by a factor of 4 but would simultaneously increase the required computer memory by a factor of 8 and the required volume of computations by a factor of 16, taking into account the decrease in the time step in the explicit scheme.

The density contours at the initial time (Fig. 12.5) and at the time of maximum radiation (Fig. 12.6) provide useful information. The radiation maximum (Fig. 12.3) is reached when the two neutron

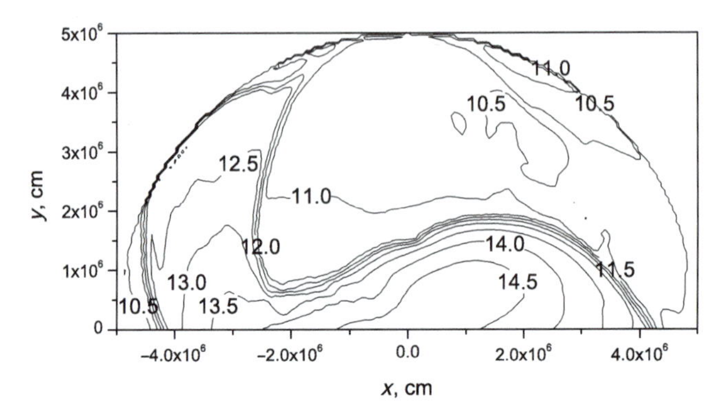

Figure 12.6 Contours of $\lg \rho$ in the equatorial plane $z = 0$ for the time of the maximum radiation, $t = 1.195 \cdot 10^{-2}$ s [15].

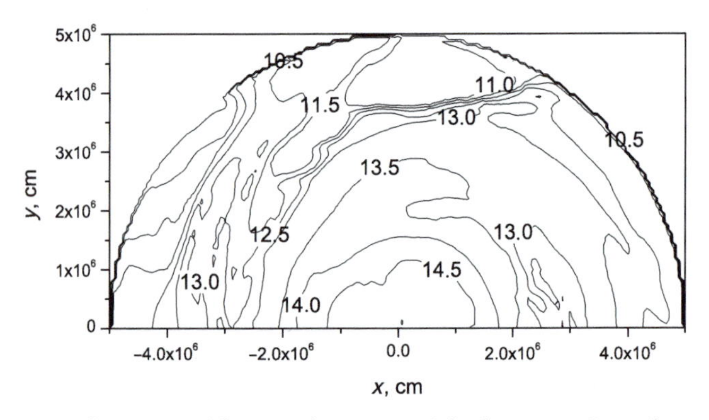

Figure 12.7 Contours of $\lg \rho$ in the equatorial plane $z = 0$ at the end of the computations, $t = 1.431 \cdot 10^{-2}$ s [15].

stars come into contact (Fig. 12.6). We attempted to reproduce these parameters in a simple model. The point-mass approximation is valid until a certain time (when the real stars come into contact, see Fig. 12.6), but we are also interested in the solution at somewhat later times (see Fig. 12.7). Therefore, in a simple model, we can *approximate the point masses by homogeneous spheres* of mass $M_\odot$. The radius of each sphere a can be estimated by equating its gravitational energy to the total energy (half of the gravitational energy) of a real neutron star with mass $M_\odot$; this yields the realistic value $4.1 \cdot 10^6$ cm. However, this radius appropriate for a single star is too large in our case. For the binary components, we can choose the lower value of $1.6 \cdot 10^6$ cm from Fig. 12.1.

The *potential energy of the system of homogeneous spheres* (with the distance between their centers x) can be integrated analytically:

$$E(x) = -GM^2 \begin{cases} (-x^5 + 30a^2x^3 - 80a^3x^2 + 192a^5)/160, & x < 2a, \\ 1/x, & 2a < x. \end{cases} \tag{12.35}$$

The computations of the gravitational-radiation back reaction and of the inertia tensor with zero trace yield exactly the same expressions as for the point masses. An additional term in the acceleration of particle i, $\ddot{x}^i_{1,2}$, corresponding to the gravitational radiation back reaction, is equal to

$$-\frac{1}{5}\frac{G}{c^5}2^6\Omega^4 M((x^i_1)^2 + (x^i_2)^2)v^i_{1,2}. \tag{12.36}$$

The equations of motion can be integrated numerically. The solution results for the gravitational waves in this case are shown in Figs. 12.8 and 12.9. There is good agreement for the initial radiation obtained in the 3D computations and the simple model. There is also good agreement for the maximum amplitudes at the moment of coalescence. Finally, the smaller number of revolutions before the maximum in the numerical computation, compared to the simplified model, is a result of numerical viscosity due to the grid resolution. This is also the origin of the underestimation of the total radiated energy in 3D computation ($\gtrsim 10^{52}$ erg). As we assumed, the grid is adequate for the radiation close to the maximum.

We should note one more interesting fact. We obtained a good agreement between the frequencies of the perturbation of the metric obtained in the 3D computations and in a simple model with spheres ($\Omega = 3.5 \cdot 10^3$ s^{-1}, Fig. 12.8); at the same time, the formula for the orbital angular velocity for point masses without the gravitational-radiation back reaction provides a lower frequency ($\Omega = 2.8 \cdot 10^3$ s^{-1}) at the time when spheres with radii $1.6 \cdot 10^6$ cm come into contact. Before the contact, the simple model is equivalent to the point-mass model.

Another interesting feature is the faster growth of the frequency with time in the model with a realistic equation of state, compared to the point-mass model [197]. Investigation of this effect requires a fine grid in 3D space. The influence of the equation of state on the

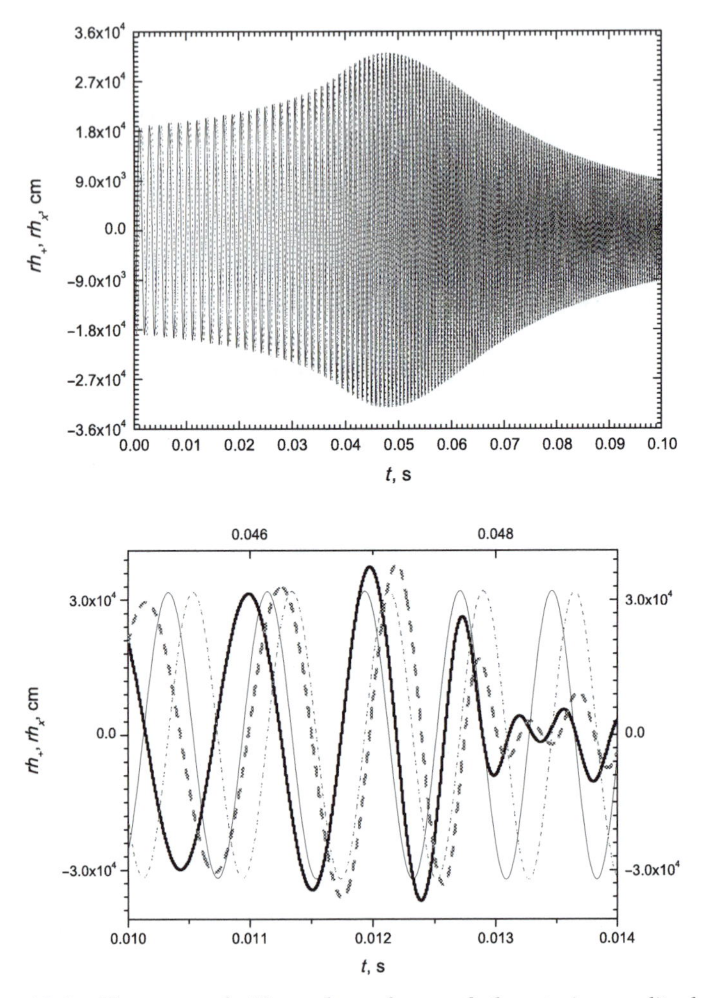

Figure 12.8 Upper panel: Time dependence of the strain amplitudes rh_+ (solid) and rh_x (dashed), as observed on the rotation axis in a simple model with homogeneous spheres. Lower panel: Comparison of the radiation close to the maximum in a 3D model (bold curves, lower x-axis) and in a simple model with spheres (thin curves, upper x). The meaning of the solid and dashed lines is the same as in the upper panel [15].

gravitational signal was first studied in Ref. [197], where a sequence of equilibrium configurations was constructed. Since the gravitational radiation was taken into account by varying the energy of the binary, the results of Ref. [197] are valid until the stars come into

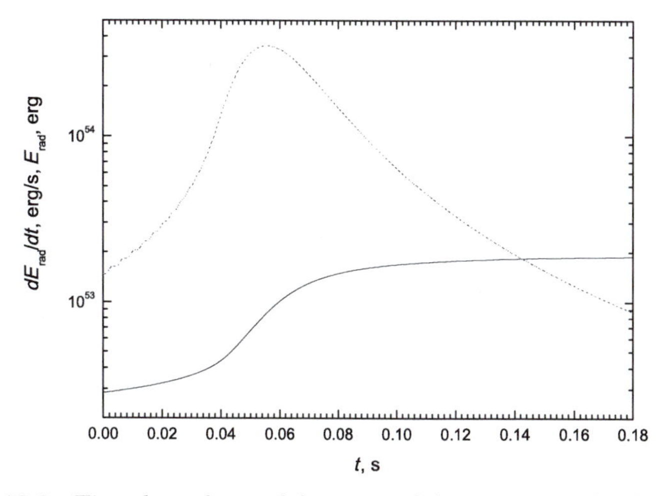

Figure 12.9 Time dependence of the power of the gravitational radiation for a simple model with homogeneous spheres [15].

contact, strictly speaking, before the maximum of the gravitational radiation. Our evolutionary gas-dynamical computations show that the radiation is maximum at the time of contact, i.e., they provide a basis for the conclusions of Ref. [197] (where the radiation maximum is present in the computational results).

12.4 Conclusion

It is interesting to compare computational results for the mergers of neutron stars with solution for the fragmentation of a rapidly rotating neutron star of the same total mass $2M_{\odot}$ [9]. At the end of the computations, we obtained a new neutron star, with some of the initial mass being ejected. The parameters of the signal for coalescence are much more favorable for observations [149]. Note that our results agree with those obtained in Refs. [270–272]. The method presented here can be applied to study the influence of the equation of state and the inner structure of the star on the parameters of the gravitational signal.

Chapter 13

Aspherical Nucleosynthesis in a Core-Collapse Supernova

The problem of nucleosynthesis was studied within an aspherical supernova model. The explosive burning was computed in a star of $25M_{\odot}$ initial mass on its final stage of evolution. The chemical composition of a presupernova was taken from realistic evolutionary computations. A piecewise parabolic method on a local stencil was applied to simulate the hydrodynamics of the explosion. The gravity was recomputed by a Poisson solver on a fine grid as the explosion developed. A detailed yield of chemical elements was performed as a post-processing step using the tracer particle method. The produced nuclei formed a layer-like structure enclosing large fragments of nickel and iron group isotopes that were pushed away from the central region by an explosion along the polar direction. The light nuclei were preferentially moving along the equatorial plane forming a torus-like structure.

13.1 Introduction

We investigated the death of a population I $25M_{\odot}$ star as a supernova (SN). According to standard evolution theory, after leaving the main sequence, this star passed the stages of a blue supergiant, a yellow supergiant, and a possible Cepheid and achieved a final red supergiant stage. Because of high metallicity, the stellar wind was strong and the star lost about a half of its mass during evolution. By the moment of explosion, the central core had a complicated structure with layers of different nuclei from iron group isotopes to oxygen and had the mass $\sim 8M_{\odot}$ [319]. When the star exhausted thermonuclear fuel in the iron core, it started to collapse until the moment when it stopped because of neutronization. The shock from bouncing matter ignited surrounding matter, and the star exploded as a Type II SN.

Normally, a presupernova rotates, which could result in a non-equilibrium neutronization process during collapse. This could lead to a strongly non-spherical explosion. The entropy excess appears in the central part and leads to the development of large-scale convection [101]. Rotation of the star deforms the shape of this high-entropy area, extending it along the axis of rotation in the direction of the highest pressure gradient — the direction of Archimedes force. This results in the formation of large-scale convective bubbles, which float up along the axis [307]. Magnetic fields also favor axisymmetry and lead to higher collimation of initial motion along the axis due to the additional magnetic pressure from the sides. According to the magneto-rotational mechanism, the magnetic field transforms the energy of rotation into the energy of the SN shock wave. For a dipole-like initial magnetic field, the SN explosion develops mainly along the rotational axis, forming a mildly collimated jet [228].

The jet-like SNe are observed in a number of cases, e.g., SN 2008D [285]. Its asymmetry was proven by a light polarization analysis, which was done with high accuracy because the neighbor SN 2007uy was used as a calibrator [146]. Also, early observations of 56Ni spectral lines in SNe could be explained by the non-sphericity of the explosions. Another example is a survey of ejecta in the Cassiopeia A SN remnant, which reveals strong asymmetry [165].

In this chapter, we present our results on nucleosynthesis in an *aspherical SN explosion* with a standard progenitor, studied by 2D hydrodynamic simulations and post-processing nucleosynthesis calculations. The explosive burning was computed in the progenitor with the chemical composition obtained in realistic evolutionary computations [319].

13.2 Methods

Hydrodynamic simulations were performed with our own numerical code based on the piecewise parabolic method on a local stencil

(PPML [257,259]). The key PPML procedure written on FORTRAN could be found in Ref. [308]. PPML is an improvement over the popular piecewise parabolic method (PPM) suggested in Ref. [108] for compressible flows with strong shocks. PPM was implemented in many modern hydrodynamic codes and is widely used in computational practice now. The principal difference between PPM and PPML is the way in which the interface values of the variables are computed between the adjacent cells. Instead of the 4-point interpolation procedure in the standard PPM, we have suggested evolving them with time using the conservation property of Riemann invariants. The PPML has demonstrated high accuracy on both smooth and discontinued solutions. It could be more suitable for problems where the requirement of low dissipation is crucial. For example, it was successfully implemented for direct turbulence simulation [195].

For SN simulation, we have used a cylindrical coordinate system where all the variables depend on the vertical coordinate z and the distance r to the z-axis, i.e., we assume the rotational symmetry. The sequential description of the PPML algorithm in cylindrical geometry is presented in Ref. [257]. Reflecting boundary conditions have been imposed along the cylindrical axis and the equatorial plane, allowing free outflow across the outer boundaries. In our case, we have performed simulations in one quadrant with a 3200×3200 grid resolution. The application of hydrodynamic code to the considered SN model showed artificially accelerated velocities of the matter near the cylindrical axis. This is a well-known pathological behavior "carbuncle" that could appear in a region where the shock is almost but not exactly parallel to the grid edges (e.g., [262]). The artificial dissipation was inserted according to Ref. [213] for the cure.

The simplest approach to the computation of explosive nucleosynthesis is to avoid coupling between hydrodynamics and nuclear reactions and to use the *tracer particles method* (TPM), which is the post-processing calculation of the nuclear yields. For the first time, it was applied to compute the nucleosynthesis in an axisymmetric explosion in a core-collapse SN with a $6M_{\odot}$ helium core with comparison to the observations of SN 1987A and solar system abundances [240]. It has become popular nowadays because of the simplicity and the possibility for following a very wide nuclear reactions network [93,214,215,239,241,247,254,297,298].

The tracers (or markers) are the points on the grid map-virtual particles that are passively advected by the flow without any coupling with it via gravity or inertia. Every tracer behaves like a massless particle. All of the tracers are advected with the flow individually with some velocity; to obtain a new location of the tracer at the next moment $t + \tau$, we have used a second-order *Runge–Kutta* time integration procedure. With the tracers we can follow the fluid elements and record the current density and temperature. These recorded values were used in the set of equations for nuclear abundances.

We have distributed 240^2 tracers over the grid in a random way at the initial moment. According to the divergence study of TPM, a number of 128 tracers per axis gives an accuracy better than 2% for nuclei with mass fraction $\gtrsim 10^{-5}$ for the grid resolution 512×512 in one hemisphere [276].

In a common approach, the mass that corresponds to each tracer is just the total mass inside the computational domain divided by the number of tracers. Thus, all the tracers represent the same mass fraction. This mass fraction is a contribution of each tracer to the total mass of produced nuclei. We suppose that this approach is not precise because the tracers are distributed in regions with different densities, which could change by several orders, and they cannot represent the same mass. In our computations, we modified standard TPM and used tracers just to reconstruct a spatial distribution of each nuclide (its mass fraction) at the final moment of hydrodynamic simulation using the triangulation technique. The mass of the nuclei produced in each computational cell was computed by multiplication of the cell's volume by the density and by the mass fraction of a particular nuclide in the center of the cell. We are going to discuss the difference between the two approaches for TPM in a future work.

We have implemented a simplified *network of nuclear reactions* from ^{4}He up to ^{56}Ni (Fig. 13.1). These are basic nuclei connected by all possible two-particle reactions with protons and α-particles in the input and output channels (the α network). Also, we have included heavy-ion reactions

$$^{12}\mathrm{C} + {}^{12}\mathrm{C} \rightarrow {}^{20}\mathrm{Ne} + \alpha,$$

$$^{12}\mathrm{C} + {}^{12}\mathrm{C} \rightarrow {}^{23}\mathrm{Na} + p,$$

$$^{12}\mathrm{C} + {}^{16}\mathrm{O} \rightarrow {}^{24}\mathrm{Mg} + \alpha,$$

$$^{12}\mathrm{C} + {}^{16}\mathrm{O} \rightarrow {}^{27}\mathrm{Al} + p,$$

$$^{16}\text{O} + {}^{16}\text{O} \rightarrow {}^{28}\text{Si} + \alpha,$$
$$^{16}\text{O} + {}^{16}\text{O} \rightarrow {}^{31}\text{P} + p, \tag{13.1}$$

and the reverse reactions due to photodisintegration. The heavy-ion reactions are shown in Fig. 13.1 by the arcs. The triple α-reaction for carbon production was also implemented,

$$3^4\text{He} \rightarrow {}^{12}\text{C} + \gamma + 7.281 \text{ MeV}. \tag{13.2}$$

In total, we have introduced 30 nuclei, which seems to be the minimum quantity required for explosive nucleosynthesis calculations with a good accuracy. We did not include nuclei heavier than ^{56}Ni since beyond the iron pike the reverse reactions are much faster than the direct ones. According to the results in other researches, production of other Ni isotopes is much less than that of ^{56}Ni (see, e.g., Ref. [215]).

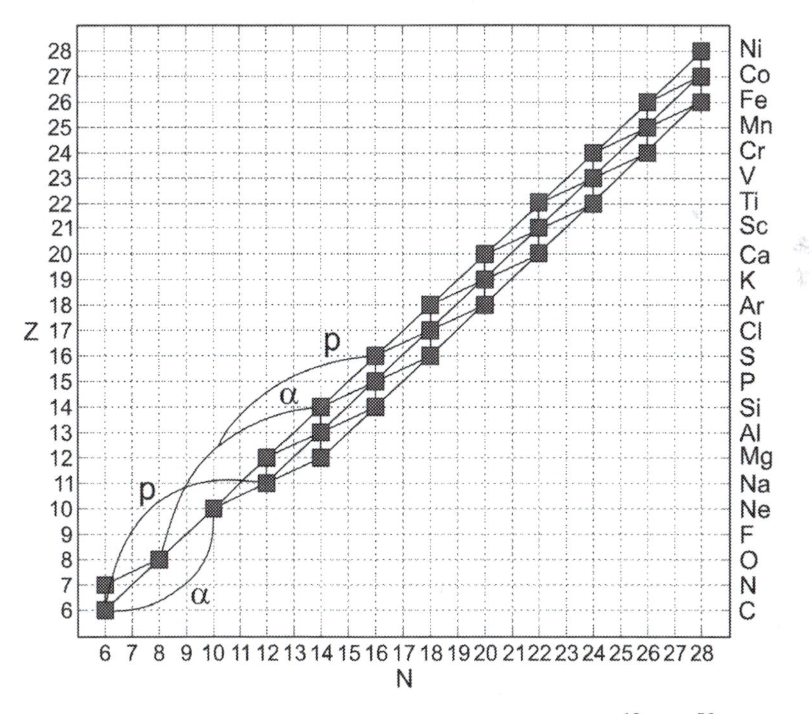

Figure 13.1 Simplified network of nuclear reactions from ^{12}C to ^{56}Ni [258].

To compute the change of nuclear abundances with time, a system of ordinary differential equations was solved [204]:

$$\frac{dY_i}{dt} = \rho Y_k Y_l R_{kl,i} - \rho Y_k Y_l R_{il,m} + Y_n \lambda_{n,i} - Y_i \lambda_{i,k}, \tag{13.3}$$

where $Y_i(t) = X_i(t)/A_i$, with X_i and A_i the mass fraction and the molar mass of nuclide i, respectively. All of the subscripts denote that sort of the nuclide, $R_{kl,i}$, the rate of the reaction $k + l \to i + \ldots$, $\lambda_{n,i}$ — probability of photodisintegration $n \to i + ^4$He per second. The timescales of the processes can differ by several orders. The RADAU5 solver for stiff problems, based on fifth-order Runge–Kutta method, was used in this case [155]. All the necessary reaction rates, based on experimental information, can be found in Refs. [92,265].

13.3 SN model

For our SN model, we have used a polytropic model of a star with index $\gamma = 4/3$, which corresponds to the Eddington standard model with purely radiative heat transport. The pressure in this case is determined by radiation. The initial configuration was calculated from the condition of hydrostatic equilibrium of a non-rotating ideal gas sphere with a simple equation of state $P = \rho RT/\mu$, where the mean molecular weight was assumed as $\mu = 0.7$.

To construct initial equilibrium profiles of density and temperature, we took central values $\rho_c \sim 4.5 \cdot 10^5$ g cm^{-3} and $T_c \sim 1 \cdot 10^9$ K that corresponded to the realistic presupernova star model [319]. This model was computed by the KEPLER code, which is a 1D implicit hydrodynamics package adopted for stellar evolution simulations [313]. The input parameters for the KEPLER code were the solar abundances.

In our *presupernova model*, we have not set the exact stellar structure according to Figure 9 from [319] but are very close to it. Upon the equilibrium profile of the density, three main regions have been set: an iron core, a layer of oxygen, and a layer of helium (Fig. 13.2). The size of each region has been chosen to give the correct mass of each type of the matter. Since we approximate the star by the polytropic equation of state, the density profile in our equilibrium configuration does not perfectly match the correct one obtained in

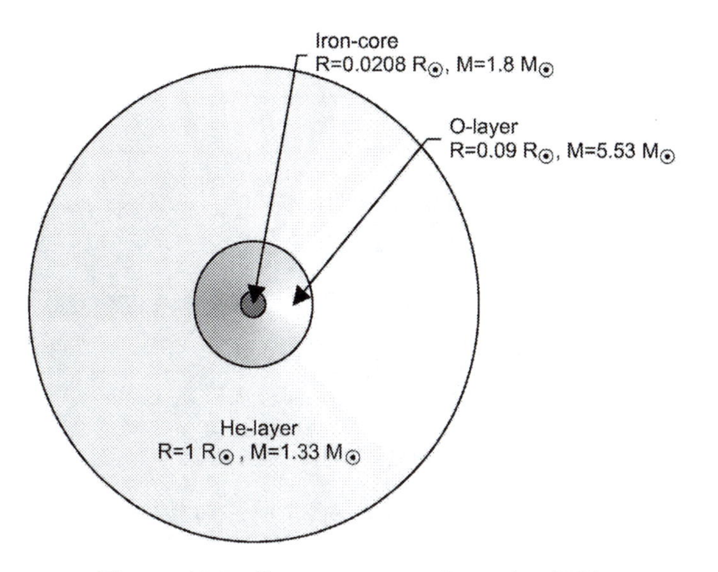

Figure 13.2 Presupernova configuration [258].

realistic evolutionary computations. To obtain the correct mass of the iron core, we assume its radius as $0.0208R_\odot$ and not $0.014R_\odot$ as in Ref. [319]. We assume that this difference is negligible for our simulation because the total size of the computational domain is $3.2 \times 3.2R_\odot$, and the most important one here is the correct gravity force.

For computational reasons beyond the helium layer, we have joined the rarefied helium–hydrogen envelope with uniform density $\rho \sim 1 \cdot 10^{-4}$ g cm^{-3}. The total initial mass inside the computational domain is $\sim$8.66 $M_\odot$. In the chemical composition, the iron group isotopes have been substituted with only one ^{54}Fe. This approximation does not make a significant impact on the results. The initial mass fractions have been set according to Ref. [319] as follows:

For the iron core ($r < 0.0208\,R_\odot$, $M_{\rm iron} = 1.8\,M_\odot$):

$$X(^{12}\mathrm{C}) = 0.03,$$

$$X(^{54}\mathrm{Fe}) = 0.97.$$

For the oxygen layer ($0.0208\,R_\odot < r < 0.09\,R_\odot$, $M_{\rm O} = 5.53\,M_\odot$):

$$X(^{20}\mathrm{Ne}) = 0.2,$$

$$X(^{24}\mathrm{Mg}) = 0.04,$$

$$X(^{12}\mathrm{C}) = 0.015,$$
$$X(^{54}\mathrm{Fe}) = 2 \cdot 10^{-4},$$
$$X(^{16}\mathrm{O}) \text{ — the rest.}$$

For the helium layer ($0.09\, R_\odot < r < 0.98234\, R_\odot$, $M_{\mathrm{He}} = 1.33\, M_\odot$):

$$X(^{12}\mathrm{C}) = 1.05 \cdot 10^{-2},$$
$$X(^{20}\mathrm{Ne}) = 1.1 \cdot 10^{-2},$$
$$X(^{54}\mathrm{Fe}) = 3 \cdot 10^{-4},$$
$$X(^{4}\mathrm{He}) \text{ — the rest.}$$

For the helium–hydrogen envelope ($r > 0.98234\, R_\odot$, $M_{\mathrm{He+H}} \sim 0.002\, M_\odot$):

$$X(^{1}\mathrm{H}) = 0.5,$$
$$X(^{14}\mathrm{N}) = 7.5 \cdot 10^{-3},$$
$$X(^{16}\mathrm{O}) = 5.1 \cdot 10^{-3},$$
$$X(^{20}\mathrm{Ne}) = 1.8 \cdot 10^{-3},$$
$$X(^{54}\mathrm{Fe}) = 3 \cdot 10^{-4},$$
$$X(^{4}\mathrm{He}) \text{ — the rest.}$$

A common way in simulations to initiate an explosion is to deposit energy to the central region of the star [320]. As a parameter, the size of the region, where the energy is deposited, should be defined. In our case, the energy $E \sim 5 \cdot 10^{51}$ erg (which is typical for SNe) is deposited in the layer of oxygen, surrounding the iron core, starting from radius $0.0208\, R_\odot$ to $0.026\, R_\odot$. The mass of this layer is $1\, M_\odot$, and this is a free parameter of the model. The iron core is excluded from hydrodynamic simulation since we assume that it collapsed and acts only as a gravity source.

The energy is divided between thermal and kinetic parts equally. In order to take into account the rotation of the star, the kinetic energy is distributed in an axisymmetric way by imposing different initial velocities in different directions: $v_z = \alpha z$ in the vertical direction and $v_r = \beta r$ on the equatorial plane. The ratio of the coefficients

$\beta : \alpha = 8 : 1$. This approach is similar to the one that was used in Ref. [215].

The tracers have been distributed randomly at the initial moment but were distributed non-uniformly because of a significant difference in density for the different regions. In the exploding layer, 7,200 tracers have been inserted; the tracers in the rest of the oxygen layer are 14,400, in the helium layer are 21,600, and in the helium–hydrogen envelope are 14,400.

A grid of 3200×3200 cells allows us to resolve the boundary between the iron core and oxygen layer and to separate accurately the collapsed core from the ejecta.

13.4 Results

The obtained structure of the hydrodynamic flow contains the contact discontinuity between the undisturbed envelope and the expanding hot matter. The contact discontinuity is followed by the shock. The narrow region between the shock and the contact discontinuity has the highest temperature $T \sim 1.5 \cdot 10^{10}$ K. Behind the shock, the flow moves more uniformly with lower temperature $T \sim (0.4 - 1) \cdot 10^9$ K. The coldest part is an inner structure with $T \sim (0.35 - 1) \cdot 10^8$ K. This large-scale hydrodynamic picture is in agreement with self-similar analytical solutions of the problem of uniformly expanding gas in a stationary medium if the profiles of density in both mediums are changing as a power law [103]. In the limiting case when the explosion could be considered as point-like, the gravitational energy is much less than the expanding kinetic energy, and the pressure of the undisturbed medium is much less than the pressure inside, the solution tends toward a Sedov–Taylor self-similar solution obtained analytically for the first time in Ref. [275]. Details of calculations can be found in the publication [258]. The details of nucleosynthesis are summarized in Fig. 13.3, where the main zones are presented on a single plot. The colors distinguish different regions according to the location in the initial presupernova configuration: The matter from the energy deposit region is colored red, the remaining part of the oxygen core is green, and the helium envelope is blue. The fragments of ^{56}Ni are pushed away by the explosion in the axial direction with the estimated velocity $v_z \sim (1.5 \ldots 2.0) \cdot 10^9$ cm s^{-1}, while the

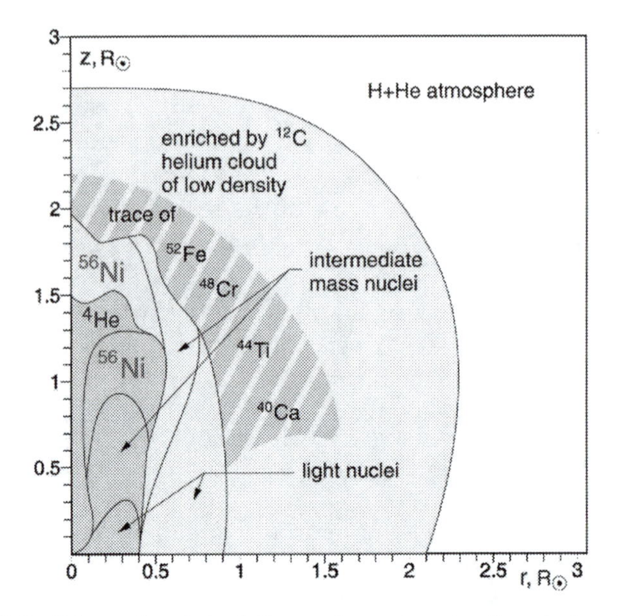

Figure 13.3 Schematic plot of the main regions in the ejecta after explosion [258].

Table 13.1 Detailed nucleosynthesis yields.

Species	Yield, $M_\odot$	Species	Yield, $M_\odot$
^{12}C	$\mathbf{6.45 \cdot 10^{-2}}$	^{38}Ar	$1.87 \cdot 10^{-5}$
^{13}N*	$4.27 \cdot 10^{-5}$	^{39}K	$4.05 \cdot 10^{-4}$
^{16}O*	**3.31**	^{40}Ca	$\mathbf{6.23 \cdot 10^{-2}}$
^{20}Ne	**0.710**	^{42}Ca	$5.91 \cdot 10^{-5}$
^{23}Na	$1.28 \cdot 10^{-6}$	^{43}Sc*	$5.75 \cdot 10^{-5}$
^{24}Mg	**0.245**	^{44}Ti*	$\mathbf{1.73 \cdot 10^{-3}}$
^{27}Al	$5.90 \cdot 10^{-5}$	^{46}Ti*	$4.86 \cdot 10^{-5}$
^{28}Si	**0.172**	^{48}Cr*	$\mathbf{4.86 \cdot 10^{-3}}$
^{30}Si	$1.23 \cdot 10^{-5}$	^{50}Cr*	$2.60 \cdot 10^{-5}$
^{31}P	$2.02 \cdot 10^{-5}$	^{51}Mn*	$4.43 \cdot 10^{-4}$
^{32}S	**0.118**	^{52}Fe*	$\mathbf{1.19 \cdot 10^{-2}}$
^{34}S	$2.76 \cdot 10^{-5}$	^{54}Fe	$1.60 \cdot 10^{-3}$
^{35}Cl	$9.03 \cdot 10^{-5}$	^{55}Co*	$1.95 \cdot 10^{-3}$
^{36}Ar	$\mathbf{3.71 \cdot 10^{-2}}$	^{56}Ni*	**0.521**

Notes: The nuclei, corresponding to the main diagonal of the α-chain, are shown in bold. The radioactive nuclei are marked with asterisks.

light nuclei are carried by the flow along the equatorial plane with $v_r \sim (2\ldots3)\cdot10^8$ cm s^{-1}. The light nuclei should form a torus-like structure around a bipolar jet of iron group isotopes. This result fully corresponds to the model of SN 2008D suggested in Ref. [290] — bipolar explosion with a torus-like distribution of oxygen.

The detailed yields obtained by TPM in our SN model are presented in Table 13.1.

13.5 Conclusion

It is interesting to compare our results to the results of Ref. [215], where a hypernova model was investigated in an axisymmetric explosion. At the moment of explosion, a progenitor star had the mass $\sim 13.8\,M_\odot$ and contained an uncollapsed layer of carbon–oxygen and a silicon core. The mass of a corresponding *main-sequence star* is $\sim 40\,M_\odot$. The burning mainly in the carbon–oxygen matter was computed by TPM with a wide network of nuclear reactions including 222 isotopes. The inserted energy of $\sim 10^{52}$ erg (twice more than in our case) was distributed in the central core (but not in a layer) with the same axial symmetry in kinetic energy. In Ref. [215], the quantity of ^{56}Ni $\sim 0.4\,M_\odot$ was a criterion to set a mass cut to reproduce the *peak of the light curve of SN* 1998bw [242]. Taking into account the difference of the progenitors, in the energies of the explosions and in the way of the energy distribution, we can say that the total masses of the main nuclei (along the diagonal of the α-chain) obtained in our SN model are reliable. The masses of the other nuclei seem to be underestimated because our network of nuclear reactions is not so rich.

The radiation from decay of *radioactive* ^{56}Ni contributes to the SN luminosity and a spectral peak. The mass of ^{56}Ni produced in SNe could be estimated from observations by tail phase luminosity analysis using a number of methods, for example, see Ref. [158]. The fainter luminosity corresponds to a smaller amount of ^{56}Ni. The summarized data on the known Type II-P SN (with plateau light curves) progenitors, including SN 1987A, show that the mass of the produced 56Ni does not exceed a value of $0.1\,M_\odot$ [283]. Still in some events more ^{56}Ni production is expected by theoretical modelizations, for example, $0.55 \pm 0.05\,M_\odot$ for SN 2003lw [218], $0.4 \pm 0.125\,M_\odot$ for

SN 2003dh [119], and $0.43 \pm 0.05\, M_\odot$ for SN 1998bw [242]. All of these are extremely powerful SNe associated with gamma-ray bursts. The estimated progenitor masses for them are 45 ± 5, 32.5 ± 7.5, and $40 \pm 5\, M_\odot$, respectively.

According to the observations, it seems that we overestimated the mass of ^{56}Ni $\sim 0.4\, M_\odot$ for this type of SN progenitor. This could be fitted by adjusting the free parameters of the model — the explosion energy and the size of the layer where it is inserted. Also, it is important to obtain asymmetry in a self-consistent way by considering the rotation. Since the size of the iron core is much smaller than the computational domain, it is better to use nested grids to resolve the core.

Chapter 14

Thermalization in Relativistic Plasma

In this chapter, we consider an application of developed numerical methods for kinetic Boltzmann equations for distribution functions of photons, electrons, and positrons interacting electromagnetically. We calculate the pair wind from the surface of strange star. This star can radiate non-equilibrium pairs in width temperature range $T < 10^{11}$ K, see Fig. 14.1. The wind is optically thin or thick for photons depending from the total luminosity. As a result, we obtained the spectrum of photons as a function of luminosity from a star. At low luminosity, one expects annihilation line 511 keV, while high luminosity case should give black body radiation obtained in 2p and 3p electromagnetic interactions.

14.1 A compact quark star

The idea that the ground state of nuclear matter is *quarks* in a deconfined state was proposed in Refs. [81,316], for review, see Ref. [314]. It was realized [31] that such a ground state may be relevant in astrophysics and compact objects referred to as quark stars [31] may be composed of such strange quark matter. Such hypothetical objects are described within the standard *Tolman–Oppenheimer–Volkoff equation* with the *equation of state* provided by the *MIT bag model* [105] $P = (\rho - 4B)/3$, where B is a constant. They may represent alternative to *neutron stars* with the following differences. Their mass–radius relation is opposite compared to the case of neutron stars: With increasing radius, the mass increases as well.

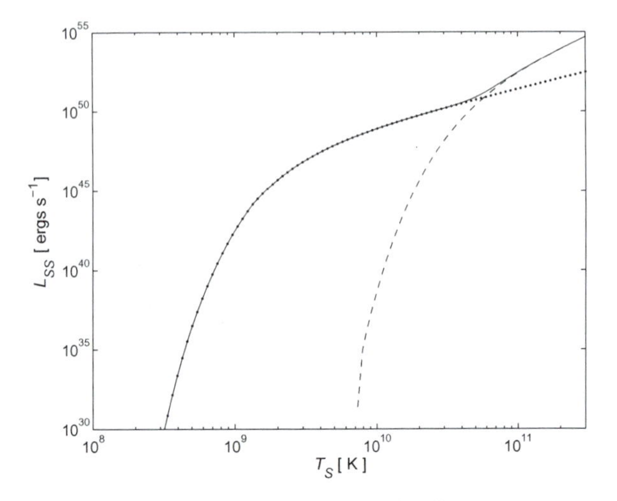

Figure 14.1 The total thermal luminosity of a hot bare strange star (solid line) as a function of the surface temperature T_S, which is the sum of the luminosity in e^+e^- pairs (dotted line) and the luminosity in thermal equilibrium photons (dashed line) [27].

Maximum masses and radii of neutron and quark stars are similar. Unlike neutron stars, the mass of the crust of quark stars is only $M_{crust} \sim 10^{-5} M_{\odot}$ [140].

The surface of quark stars is composed of deconfined quarks in *statistical equilibrium*, subject to short range nuclear forces. At the same time, electrons obey long-range electromagnetic interaction and their spatial distribution does not coincide with quarks. This fact results in the presence of strong electric fields, referred to as *electrosphere* at the surface of quark stars [306].

If the surface of quark stars is exposed to the vacuum, it is called a bare quark star [314]. The electric field on the surface of bare quark star exceeds the Schwinger *critical electric field* for pair production [302]. At absolute zero temperature, pair production is blocked by the Pauli principle. If the temperature increases, the star can produce a wind of electron–positron pairs, but due to large density, its radiation in photons is suppressed [31] for temperatures below 10^{10} K.

For the typical radius of a quark star (10^6 km), a temperature of $T_S \simeq 10^9$–10^{10} K gives an energy injection rate in pairs $\dot{E} \simeq 10^{43}$–10^{49} ergs s^{-1} [305]. For such powerful winds, the pair density near the surface is very high, the wind is opaque for photons and the pairs and photons are in *thermal equilibrium* almost up to the wind photosphere. The outflow may then be described fairly well by relativistic hydrodynamics [145,148,252,253]. The emerging emission consists mostly of photons, so $L_\gamma \simeq \dot{E}$. The photon spectrum is roughly Planckian with a temperature of $\sim 10^{10}(\dot{E}/10^{49}\ \text{ergs s}^{-1})^{1/4}$ K. The emerging luminosity in $e^\pm$ pairs is very small, $L_e = \dot{E} - L_\gamma \simeq 10^{-7}(\dot{E}/10^{49}\ \text{ergs s}^{-1})^{-1/4} L_\gamma$. All this applies roughly down to $\dot{E} \sim 10^{42}$ ergs s^{-1}.

14.2 The pair plasma wind from the compact quark star

In contrast, for $\dot{E} < 10^{42}$ ergs s^{-1} ($T_\text{S} < 9 \times 10^8$ K), the thermalization time for the pairs and photons is longer than the escape time $t_\text{esc} \simeq$ a few times R/c, and pairs and photons are not in thermal equilibrium. The results of numerical calculations of the characteristics of the emerging emission in pairs and photons in stationary winds with energy injection rates $\dot{E} = 10^{35} - 10^{42}$ ergs s^{-1} are reported in the following.

Consider an *electron–positron pair wind* that flows away from a hot, bare, unmagnetized, quark star with a radius of $R = 10^6$ cm. Pairs are injected from time $t = 0$, at a constant rate into the wind, which is assumed spherical. This result in a time-dependent wind that becomes stationary after a time $\sim$100 ms.

The relativistic Boltzmann equations for the pairs and photons (2.18) are solved numerically in Refs. [25,26] within special relativity and in Ref. [27] within general relativity.

The non-stationary pair plasma wind in the region outside the *quark star* is considered if the quark star is considered as an internal boundary with the constant energy flux of pairs, so one expects to obtain the steady solution at sufficiently large times. At the internal boundary, $r = R$, the input pair number flux depends on the temperature T_S at the stellar surface alone and is taken as [305]

$$\dot{F}_\pm = 10^{39}\left(\frac{T_\text{S}}{10^9\,\text{K}}\right)^3 \exp\left[-11.9\left(\frac{T_\text{S}}{10^9\,\text{K}}\right)^{-1}\right] \times\left[\frac{\zeta^3\ln(1+2\zeta^{-1})}{3(1+0.074\zeta)^3} + \frac{\pi^5\zeta^4}{6(13.9+\zeta)^4}\right]\text{cm}^{-2}\text{s}^{-1}, \quad (14.1)$$

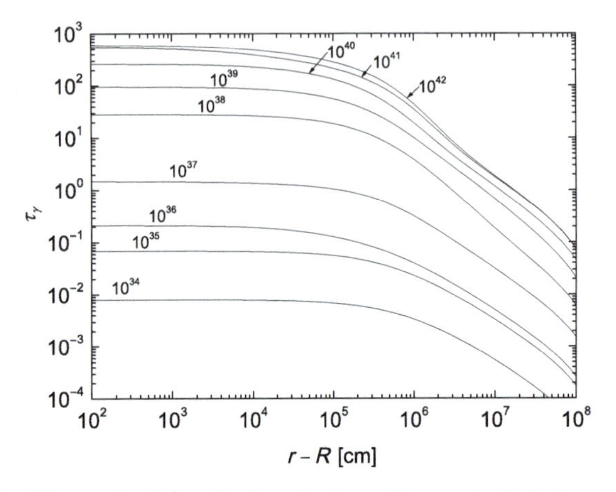

Figure 14.2 The optical depths for photons, from r to infinity, as functions of the distance from the stellar surface for different values of $\dot{E}$, as marked on the curves [26].

where $\zeta = 20(T_{\rm S}/10^9\,{\rm K})^{-1}$. Their energy spectrum is thermal with temperature $T_{\rm S}$, and their angular distribution is isotropic. The energy injection rate in $e^\pm$ pairs is then

$$\dot{E} = 4\pi R^2[m_e c^2 + (3/2)k_{\rm B}T_{\rm S}]\dot{F}_\pm\,. \qquad (14.2)$$

For the range of the energy injection rate, we consider $\dot{E} \leq 10^{42}$ erg s^{-1} and $T_{\rm S}$ is $\lesssim 10^9$ K (see Fig. 14.2).

The surface of the *quark star* is assumed to be a perfect mirror for both $e^\pm$ pairs and photons. At the external boundary ($r = r_{\rm ext}$), the pairs and photons escape freely.

Figure 14.3 shows the particle number fluxes in $e^\pm$ pairs ($\dot{N}_e$) and photons ($\dot{N}_\gamma$) as functions of the distance from the stellar surface. At $r = R$, the particle number flux of injected pairs is

$$\dot{N}_e^{\rm in} \simeq 10^{48}(\dot{E}/10^{42}\,{\rm ergs\ s^{-1}})\ {\rm s}^{-1}, \qquad (14.3)$$

while the photon number flux is zero ($\dot{N}_\gamma^{\rm in} = 0$). There is the upper limit on the emerging flux of pairs that is $\dot{N}_e^{\rm max} \simeq 10^{43}$ s^{-1}. If $\dot{N}_e^{\rm in} \gg \dot{N}_e^{\rm max}$, the pair flux $\dot{N}_e$ decreases strongly at the distance

$$l_{\rm ann} \simeq (n_e\sigma_{\rm ann})^{-1} \simeq 10^{-2}(\dot{E}/10^{42}\,{\rm ergs\ s^{-1}})^{-1}\ {\rm cm} \qquad (14.4)$$

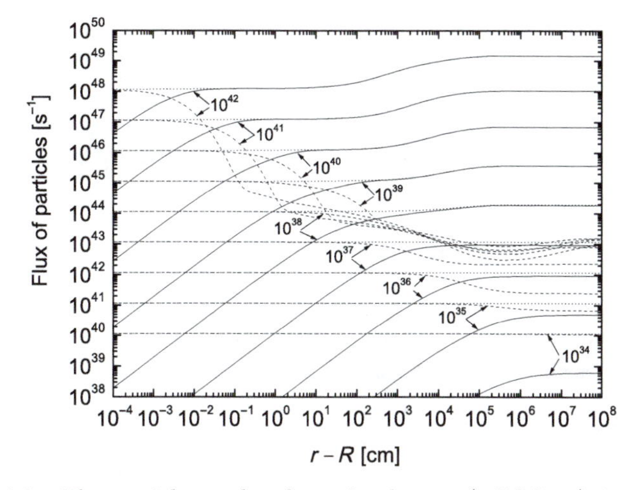

Figure 14.3 The particle number fluxes in photons (solid lines), in e^+e^- pairs (dashed lines), and the total (dotted lines) as functions of the distance from the stellar surface, for different values of $\dot{E}$, as marked on the curves for photons. Very near the stellar surface, $r - R \lesssim 10^{-4}$ cm, the particle number fluxes in pairs are equal to $\dot{N}_e^{\text{in}}$. Using this and Eq. (14.3), the curves of the pair fluxes for different values of Ė may be easy identified [26].

from the stellar surface, and the emerging flux of pairs is equal to $\dot{N}_e^{\max}$ within a factor of 2, where $n_e \simeq F_e/v_{\text{out}}$ is the density of pairs near the stellar surface, $\sigma_{\text{ann}} \simeq \sigma_{\text{T}}(c/v_e)$ is the cross-section of pair annihilation, σ_{T} is the Thomson cross-section, and $v_e \simeq (3k_{\text{B}}T_{\text{S}}/m_e)^{1/2} \simeq 10^{10}$ cm s^{-1} is the mean velocity of injected pairs. At $\dot{E} > 10^{37}$ ergs s^{-1}, the radiative processes are essential, and the total flux of particles increases in the process of the wind outflow (see Fig. 14.3). For $\dot{E} = 10^{42}$ ergs s^{-1}, when photons strongly prevail in the emerging emission, the emerging flux of photons increases by a factor of 15 in comparison with $\dot{N}_e^{\text{in}}$. The energy fluxes in $e^{\pm}$ pairs and photons vary with the distance from the stellar surface more or less similar to the particle number fluxes, except of that the total energy flux is constant in any run and equal to $\dot{E}$. The upper limit on the emerging luminosity in $e^{\pm}$ pairs is $L_e^{\max} \simeq 2 \times 10^{37}$ ergs s^{-1}; here the rest energies of the emerging pairs are included. The existence of the upper limits $\dot{N}_e^{\max}$ and $L_e^{\max}$ is connected with the production of photons via radiative three body processes. Indeed, Figs. 8 and 9 show the fluxes of particles and energy, respectively, when only two body processes are taken into account. We see that at $\dot{E} \gg 10^{37}$

Table 14.1 Physical processes included in simulations.

Basic two-body interaction	Radiative variant
Møller and Bhaba scattering	Bremsstrahlung
$ee \to ee$	$ee \leftrightarrow ee\gamma$
Compton scattering	Double Compton scattering
$\gamma e \to \gamma e$	$\gamma e \leftrightarrow \gamma e\gamma$
Pair annihilation	Three photon annihilation
$e^+e^- \to \gamma\gamma$	$e^+e^- \leftrightarrow \gamma\gamma\gamma$
Photon–photon pair production	
$\gamma\gamma \to e^+e^-$	

ergs s^{-1} and $r - R \gg l_{\rm ann}$ these fluxes in $e^{\pm}$ pairs are only an order of magnitude less than those in photons.

Although the injected pair plasma contains no radiation, as the plasma moves outward, photons are produced by pair interactions. The two-particle and three-particle interactions between pairs and photons listed in Table 14.1 are considered.

In the following, we present the results for the structure of stationary pair winds and their emerging emission. For such a wind at the $\dot{E}$ region we explore (10^{34} ergs s^{-1} $\leq \dot{E} \leq 10^{42}$ ergs s^{-1}), Fig. 14.2 shows the optical depth $\tau_\gamma(r)$ for photons, from r to infinity. The pair wind is optically thick [$\tau_\gamma(R) > 1$] at $\dot{E} > 10^{37}$ ergs s^{-1}. The radius of the wind photosphere $r_{\rm ph}$ determined by condition $\tau(r_{\rm ph}) = 1$ varies from R at $\dot{E} = 10^{37}$ ergs s^{-1} to $\sim 20R = 2 \times 10^7$ cm at $\dot{E} = 10^{42}$ ergs s^{-1}. The wind photosphere is deeply inside the external boundary ($r_{\rm ph} \lesssim 0.1 r_{\rm ext}$), and the inward ($\mu < 0$) flux of particles at $r = r_{\rm ext}$ may be neglected.

The results for the properties of the emerging radiation after stationarity is achieved are given, so the total wind luminosity is equal to the energy injection rate: $L = L_e + L_\gamma = \dot{E}$. The results are presented for different values of L, which is the only free parameter. The corresponding surface temperature $T_{\rm S}$ is found from Eqs. (14.1) and (14.2). For $L > L_* \simeq 2 \times 10^{35}$ ergs s^{-1}, the emerging emission consists mostly of photons, while pairs dominate for $L < L_*$.

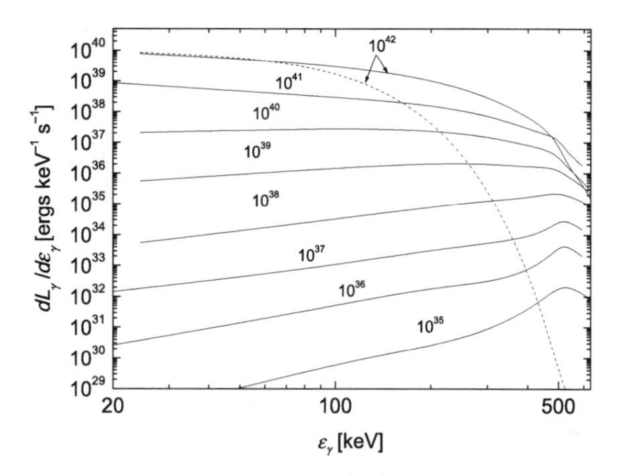

Figure 14.4 The energy spectra of emerging photons for different values of the total luminosity (measured in erg s^{-1}), as marked on the curves. Also shown, for $L = 10^{42}$ erg s^{-1} sec, is the *Planck spectrum* with the same energy density as that of the photons at the photosphere of the outflowing wind (dashed line). This figure is based on improved calculations with correct *Jacobi matrix* important for thermalization. The original calculations were presented in Ref. [26].

This simply reflects the fact that for $L < L_*$ the pair annihilation time is longer than the escape time, and the injected pairs remain mostly intact. At higher luminosities, most pairs annihilate before escape, and reconversion into pairs is inefficient, as the mean energy of photons at the photosphere is rather below pair-creation threshold.

At low luminosities, $L \sim 10^{35}$–10^{37} ergs s^{-1}, photons formed in the wind escape relatively freely, and the photon spectra resemble a very wide annihilation line. The small decrease in mean photon energy $\langle\epsilon_\gamma\rangle$ from ~500 keV at $L \simeq 10^{35}$ ergs s^{-1} to ~400 keV at $L \simeq 10^{37}$ ergs s^{-1} occurs because of the energy transfer from annihilation photons to pairs via Compton scattering, see Fig. 14.4. As a result of this transfer, the emerging pairs are heated up to the mean energy $\langle\epsilon_e\rangle \simeq 400$ keV at $L \simeq 10^{37}$ ergs s^{-1}. For $L > 10^{37}$ ergs s^{-1}, changes in the particle number due to three-body processes are essential, and their role in thermalization of the outflowing plasma increases with the increase in L. For $L = 10^{42}$ ergs s^{-1}, the photon spectrum is almost Planckian, except for the presence of a high-energy tail at $\epsilon_\gamma > 100$ keV. At this luminosity, the mean energy of the emerging photons is $\sim$40 keV, while the mean energy of the blackbody photons is $\sim$30 keV.

For $L \simeq 10^{42}$ ergs s^{-1}, the emerging pair energy spectrum is close to a Maxwellian, while for $L \lesssim 10^{40}$ ergs s^{-1}, it deviates significantly from it. All details of the steady wind for the calculated luminosities can be found in Ref. [26]. Later in Ref. [27] two important effects were considered: influence of the general relativistic effects on the photon spectrum and the admixture of small amount of non-thermal radiation from the surface of the *quark star*. It was shown that accounting for general relativity leads to redshift of the annihilation line. In addition, the admixture on the non-thermal photons does not remove the annihilation line at the luminosities 10^{38}–10^{40} erg s^{-1}.

It is appropriate to make some remarks. The annihilation line and high-energy tail in the photon spectrum can be a diagnostic tool for the *super-Eddington luminosities* $\gtrsim 10^{38}$ erg s^{-1}. These features can be affected by possible existence of some amount of usual matter around the bare *quark star*. The results reported above show that photon spectra becomes Planckian at luminosities $\gtrsim 10^{42}$ erg s^{-1}. In principle, the bare *quark star* can support very large luminosities $\gtrsim 10^{50}$ erg s^{-1}. Hence, a candidate *quark star* is a compact object with the super-Eddington luminosity.

In particular, soft γ-ray repeaters (SGRs), which are the sources of short bursts of hard X-rays with super-Eddington luminosities up to $\sim 10^{42}$–10^{45} ergs s^{-1}, are reasonable candidates for *quark stars*, see, e.g., Refs. [31,117,305]. The bursting activity of SGRs may be explained by fast heating of the stellar surface up to the temperature of $\sim(1-2)\times 10^{9}$ K and its subsequent thermal emission [303,304]. The heating mechanism may be either impacts of comets onto bare *quark stars* [321] or fast decay of superstrong ($\sim 10^{15}$ G) magnetic fields [294].

For typical luminosities of SGRs $L \sim 10^{41}$–10^{42} erg s^{-1}, the mean photon energy is $\sim$40 keV, which is consistent with observations of SGRs [164].

Another important idiosyncrasy found is a strong anti-correlation between spectral hardness and luminosity. While at very high luminosities ($L > 10^{42}$–10^{43} erg s^{-1}) the spectral temperature increases with luminosity as in thermal radiation, in the range of luminosities studied, where *thermal equilibrium* is not achieved, the expected correlation is opposite. Such anti-correlations were, indeed, observed for

SGR 1806-20 and SGR 1900+14 where the burst statistics is high enough [129,147,166].

The recent progress in the theory of *quark stars* concerns photon emission. In particular, it was confirmed that for low luminosities $L \leq 10^{40}$ erg s^{-1}, the radiation in electron–positron pairs dominates that of photons [90]. It was also shown that bremsstrahlung in strong electric field of the *electrosphere* can be comparable to the output in pairs [323]. This effect has been discussed above, for details, see Ref. [27]. For a review on astrophysical constraints on quark stars, see Ref. [299].

Appendix A

Hydrodynamic Equations in Orthogonal Curvilinear Coordinates

In orthogonal curvilinear coordinates x^i, the *metric tensor* is

$$g_{ij} = \left(\frac{\partial \mathbf{r}}{\partial x^i}, \frac{\partial \mathbf{r}}{\partial x^j}\right) = \text{diag}(g_{11}, g_{22}, g_{33}), \; g^{ij} = \text{diag}(g_{11}^{-1}, g_{22}^{-1}, g_{33}^{-1}), \tag{A.1}$$

the *Christoffel symbols* are

$$\Gamma^k_{ij} = \frac{1}{2} g^{kk}\left(\frac{\partial g_{ik}}{\partial x^j} + \frac{\partial g_{jk}}{\partial x^i} - \frac{\partial g_{ij}}{\partial x^k}\right) = \frac{1}{2g_{kk}}\left(\frac{\partial g_{ik}}{\partial x^j} + \frac{\partial g_{jk}}{\partial x^i} - \frac{\partial g_{ij}}{\partial x^k}\right), \tag{A.2}$$

and the *covariant derivative of the tensor* is

$$\nabla_l T^{i_1\cdots j_p}_{j_1\cdots j_q} = \frac{\partial T^{i_1\cdots j_p}_{j_1\cdots j_q}}{\partial x^l} + \Gamma^{i_1}_{ls} T^{s\cdots i_p}_{j_1\cdots j_q} + \Gamma^{i_p}_{ls} T^{i_1\cdots s}_{j_1\cdots j_q} - \Gamma^s_{lj_1} T^{i_1\cdots i_p}_{s\cdots j_q} - \Gamma^s_{lj_q} T^{i_1\cdots i_p}_{j_1\cdots s}. \tag{A.3}$$

One uses the norm of units vectors for Riemannian geometry.

The above definitions are enough to derive all other derivatives. The *gradient of the scalar* is

$$g^{ij}\nabla_j T = \frac{1}{g_{ii}}\frac{\partial T}{\partial x^i}, \tag{A.4}$$

the *gradient of the vector* is

$$\nabla_j u^i = \frac{\partial u^i}{\partial x^j} + \Gamma^i_{js} u^s = \frac{\partial u^i}{\partial x^j} + \frac{1}{2g_{ii}} \left(\sum_s \frac{\partial g_{ij}}{\partial x^s} u^s + \frac{\partial g_{ii}}{\partial x^j} u^i - \frac{\partial g_{jj}}{\partial x^i} u^j \right), \tag{A.5}$$

the *divergence of the vector* is

$$\begin{aligned} \nabla_i u^i &= \frac{\partial u^i}{\partial x^i} + \Gamma^k_{kj} u^j = \frac{\partial u^i}{\partial x^i} + \frac{1}{2g_{kk}} \frac{\partial g_{kk}}{\partial x^i} u^i \\ &= \frac{\partial u^i}{\partial x^i} + \sum_k \frac{\partial \ln \sqrt{g_{kk}}}{\partial x^i} u^i = \frac{1}{\sqrt{|g|}} \frac{\partial \sqrt{|g|} u^i}{\partial x^i}, \end{aligned} \tag{A.6}$$

and the *divergence of the tensor* is

$$\begin{aligned} \nabla_i \sigma^{ij} &= \sum_i \left(\frac{\partial \sigma^{ij}}{\partial x^i} + \Gamma^i_{is} \sigma^{sj} + \Gamma^j_{is} \sigma^{is} \right) \\ &= \sum_i \left(\frac{1}{\sqrt{|g|}} \frac{\partial \sqrt{|g|} \sigma^{ij}}{\partial x^i} + \frac{\partial \ln g_{jj}}{\partial x^i} \sigma^{ij} - \frac{1}{2g_{jj}} \frac{\partial g_{ii}}{\partial x^j} \sigma^{ii} \right). \end{aligned} \tag{A.7}$$

Hydrodynamical conservation laws (for the one-component ideal fluid without viscosity) are

$$\frac{\partial \rho}{\partial t} + \nabla_i (\rho u^i) = 0, \tag{A.8}$$

$$\frac{\partial \rho u^j}{\partial t} + \nabla_i \sigma^{ij} = 0, \tag{A.9}$$

and

$$\frac{\partial \rho E}{\partial t} + \nabla_i (\rho E + P) u^i = \nabla_i \left(\kappa g^{ii} \frac{\partial T}{\partial x^j} \right) + \rho Q, \tag{A.10}$$

where $E = \epsilon + u^i g_{ik} u^k / 2$, $\kappa = \kappa(\rho, T)$, $Q = Q(\rho, T)$, and $\sigma^{ij} = \rho u^i u^j + P g^{ij}$.

Let us introduce new "physical", "vectors", and "tensors" in physical units for all components:

$$v_i \equiv \sqrt{g_{ii}} u^i, \ E = \epsilon + \frac{v_i^2}{2}, \tag{A.11}$$

$$\Pi_{ij} \equiv \sqrt{g_{ii} g_{jj}} \sigma^{ij} = \rho v_i v_j + P\delta_{ij}. \tag{A.12}$$

Then

$$\nabla_m u^m = \frac{1}{\sqrt{|g|}} \sum_i \frac{\partial}{\partial x^i} \left(\frac{\sqrt{|g|} v_i}{\sqrt{g_{ii}}} \right) \equiv \mathrm{div}\mathbf{v}. \tag{A.13}$$

The *mass conservation law* is

$$\frac{\partial \rho}{\partial t} + \mathrm{div}(\rho \mathbf{v}) = 0. \tag{A.14}$$

The *momentum conservation law* is

$$\frac{\partial \rho \frac{v_j}{\sqrt{g_{jj}}}}{\partial t} + \sum_i \left(\frac{1}{\sqrt{|g|}} \frac{\partial \sqrt{|g|} \frac{\sigma_{ij}}{\sqrt{g_{ii} g_{jj}}}}{\partial x^i} + \frac{\partial \ln g_{jj}}{\partial x^i} \frac{\sigma_{ij}}{\sqrt{g_{ii} g_{jj}}} \right.$$

$$\left. - \frac{1}{2 g_{jj}} \frac{\partial g_{ii}}{\partial x^j} \frac{\sigma_{ii}}{\sqrt{g_{ii} g_{ii}}} \right) = 0 \tag{A.15}$$

or

$$\frac{\partial \rho v_j}{\partial t} + (\mathrm{Div}\Pi)_j = 0, \tag{A.16}$$

where

$$\mathrm{Div}\Pi_j \equiv \frac{1}{\sqrt{g_{jj}}} \sum_i \left(\frac{1}{\sqrt{|g|}} \frac{\partial}{\partial x^i} \left(\sqrt{\frac{|g| g_{jj}}{g_{ii}}} \Pi_{ij} \right) - \frac{\partial \ln \sqrt{g_{ii}}}{\partial x^j} \Pi_{ii} \right). \tag{A.17}$$

The *energy conservation law* is

$$\frac{\partial \rho E}{\partial t} + \mathrm{div}((\rho E + P)\mathbf{v} - \kappa \mathrm{grad} T) = \rho Q, \tag{A.18}$$

where

$$(\mathrm{grad} T)_i \equiv \frac{1}{\sqrt{g_{ii}}} \frac{\partial T}{\partial x^i}. \tag{A.19}$$

Appendix B

Collision Integrals for Weak Interactions

Here, we omit the dependence on the coordinate r and use the notation

$$n_\nu(\epsilon, \mu, t) = \frac{E_\nu(\epsilon, \mu, t)}{\epsilon}. \tag{B.1}$$

The computations were carried out on a grid in $\epsilon_{\omega+1/2}$, $\mu_{k+1/2}$. Let us now consider the computation of the absorption and emission coefficients in certain cases in more detail.

B.1 Scattering of neutrinos on electrons

Taking into account the filled final electron states, one has

$$\begin{aligned}\left(\frac{\partial f_\nu(\mathbf{q}, \mathbf{t})}{\partial t}\right)_{\nu e} = \int \frac{V d\mathbf{p} V d\mathbf{q}' V d\mathbf{p}'}{(2\pi\hbar)^6} w_{\mathbf{q}',\mathbf{p}';\mathbf{q},\mathbf{p}} \\ \times \big[(1 - F_e(\mathbf{p}, t)) f_\nu(\mathbf{q}', t) f_e(\mathbf{p}', t) \\ - (1 - F_e(\mathbf{p}', t)) f_\nu(\mathbf{q}, t) f_e(\mathbf{p}, t)\big],\end{aligned} \tag{B.2}$$

where $F_e \equiv \frac{(2\pi\hbar c)^3}{2} f_e$, the probability of the process is

$$w_{\mathbf{q}',\mathbf{p}';\mathbf{q},\mathbf{p}} = c(2\pi\hbar)^4 \frac{(\hbar c)^4}{V^3} \delta(\mathfrak{q} + \mathfrak{p} - \mathfrak{q}' - \mathfrak{p}') \frac{|M_{fi}|^2}{16\epsilon_\nu \epsilon_e \epsilon'_\nu \epsilon'_e}, \tag{B.3}$$

the square of the matrix element is [84,300]

$$|M_{fi}|^2 = 32\left(\frac{G_F}{(\hbar c)^3}c^2\right)^2 \left[(C_V + C_A)^2(\mathfrak{pq})(\mathfrak{p}'\mathfrak{q}') \right.$$
$$\left. + (C_V - C_A)^2(\mathfrak{p}'\mathfrak{q})(\mathfrak{pq}') - (C_V^2 - C_A^2)m_e^2c^2(\mathfrak{qq}')\right], \quad \text{(B.4)}$$

where $C_V = 1+\frac{a}{2m_Z^2}\frac{m_W^2}{g^2} = \frac{1}{2}+2\sin^2\theta_W = 1.2$, $C_A = 1-\frac{b}{2m_Z^2}\frac{m_W^2}{g^2} = \frac{1}{2}$, the mass of the intermediate bosons is $m_W = 37.3/\sin\theta_W$ GeV, $m_Z = 74.6/\sin(2\theta_W)$ GeV, the tangent of the Weinberg mixing angle is $\tan\theta_W = g'/g$, and the Fermi constant G_F is given by $\frac{G_F}{(\hbar c)^3} = \frac{\sqrt{2}}{8}\frac{g^2}{m_W^2c^4} = \frac{1.015\cdot 10^{-5}}{m_p^2c^4}$. In the computations, it is used as $\mathfrak{q} = \frac{\epsilon_\nu}{c}(1, \mathbf{e}_\nu)$, $\mathfrak{p} = \frac{\epsilon_e}{c}(1, \beta\mathbf{e}_e)$, $d\mathbf{q} = d\epsilon_\nu d\omega_\nu\epsilon_\nu^2/c^3$, $d\mathbf{p} = d\epsilon_e d\omega_e\epsilon_e^2\beta/c^3$, and $do = d\mu d\phi$. The conservation of energy–momentum yields the energies of the product particles as a function of their initial momenta and the final angle of the neutrino trajectory:

$$\epsilon_\nu' = \frac{\epsilon_\nu\epsilon_e(1-\beta_e\mathbf{b}_e\cdot\mathbf{b}_\nu)}{\epsilon_e(1-\beta_e\mathbf{b}_e\cdot\mathbf{b}_\nu') + \epsilon_\nu(1-\beta_e\mathbf{b}_\nu\cdot\mathbf{b}_\nu')}, \quad \text{(B.5)}$$

$$\epsilon_e' = \epsilon_e + \epsilon_\nu - \epsilon_\nu', \quad \text{(B.6)}$$

where the unit vector $\mathbf{b}_i \equiv \mathbf{p}_i/p_i$ coincides with the direction of the particle momentum i. The direction of the final electron trajectory is $\mathbf{b}_e' = (\beta_e\epsilon_e\mathbf{b}_e + \epsilon_\nu\mathbf{b}_\nu - \epsilon_\nu'\mathbf{b}_\nu')/(\beta_e'\epsilon_e')$.

Consider the absorption process in more detail. Using a delta function for the momenta, one can integrate the reaction rate over $d\mathbf{p}'$. In the following integration over $d\mathbf{q}' = d\epsilon_\nu'(\epsilon_\nu')^2d\omega_\nu'/c^3$, one should use ϵ_e' from the previous momentum conservation relation, $\mathbf{p}' = \mathbf{q}+\mathbf{p}-\mathbf{q}'$:

$$\int d\epsilon_\nu'\delta(\epsilon_\nu + \epsilon_e - \epsilon_\nu' - \epsilon_e')$$
$$= \int d(\epsilon_\nu' + \epsilon_e')\left|\frac{\partial(\epsilon_\nu' + \epsilon_e')}{\partial\epsilon_\nu'}\right|^{-1}\delta(\epsilon_\nu + \epsilon_e - \epsilon_\nu' - \epsilon_e') = |J_{\nu e}|, \quad \text{(B.7)}$$

where

$$J_{\nu e}^{-1} = \frac{\partial(\epsilon_\nu' + \epsilon_e')}{\partial\epsilon_e'} = 1 + \frac{\partial(\epsilon_e')}{\partial\epsilon_\nu'}. \quad \text{(B.8)}$$

Using the momentum conservation,

$$\beta'_e \epsilon'_e = (\epsilon_\nu \mathbf{b}_\nu + \beta_e \epsilon_e \mathbf{b}_e - \epsilon'_\nu \mathbf{b}'_\nu)\mathbf{b}'_e, \tag{B.9}$$

one obtains

$$\frac{\partial(\beta'_e \epsilon'_e)}{\partial \epsilon_\nu} = \frac{1}{\beta'_e}\frac{\partial(\epsilon'_e)}{\partial \epsilon'_\nu} = -\frac{\partial(\epsilon'_\nu)}{\partial \epsilon'_\nu}\mathbf{b}'_\nu \mathbf{b}'_e = -\mathbf{b}'_\nu \mathbf{b}'_e, \tag{B.10}$$

$$J_{\nu e} = \frac{1}{1 - \beta'_e \mathbf{b}'_\nu \mathbf{b}'_e}. \tag{B.11}$$

The absorption coefficient for the neutrino for the considered reaction is

$$\begin{aligned}\chi_\nu^{\nu e} f_\nu &= -\frac{1}{c}\left(\frac{\partial f_\nu}{\partial t}\right)^{\text{abs}}_{\nu e} \\ &= \int dn_\nu dn_e do'_\nu J_{\nu e}(1 - F_e(\mathbf{p}', t))\frac{(\epsilon'_\nu)^2 |M_{fi}|^2 \hbar^2 c^2}{16\epsilon_\nu \epsilon_e \epsilon'_\nu \epsilon'_e} f_\nu,\end{aligned} \tag{B.12}$$

where $dn_i = d\epsilon_i do_i \epsilon_i^2 \beta_i f_i / c^3 = d\epsilon_i d\omega_i E_i/(2\pi\epsilon_i)$. One can obtain the remaining numerical coefficients in similar fashion:

$$\begin{aligned}(\chi E)^{\nu e}_{\nu,\omega,k}(T, \mu_e) &= \frac{1}{\Delta\epsilon_{\nu,\omega}\Delta\mu_k}\int_{\epsilon_\nu \in \Delta\epsilon_{\nu,\omega}, \mu_\nu \in \Delta\mu_k} dn_\nu dn_e do'_\nu \\ &\quad \times J_{\nu e}(1 - F_e(\mathbf{p}', t))\frac{{\epsilon'_\nu}^2 \epsilon_\nu |M_{fi}|^2 \hbar^2 c^2}{16\epsilon_\nu \epsilon_e \epsilon'_\nu \epsilon'_e},\end{aligned} \tag{B.13}$$

$$\begin{aligned}\eta^{\nu e}_{\nu,\omega,k}(T, \mu_e) &= \frac{1}{\Delta\epsilon_{\nu,\omega}\Delta\mu_k}\int_{\epsilon'_\nu \in \Delta\epsilon_{\nu,\omega}, \mu'_\nu \in \Delta\mu_k} dn_\nu dn_e do'_\nu \\ &\quad \times J_{\nu e}(1 - F_e(\mathbf{p}', t))\frac{{\epsilon'_\nu}^2 \epsilon'_\nu |M_{fi}|^2 \hbar^2 c^2}{16\epsilon_\nu \epsilon_e \epsilon'_\nu \epsilon'_e}.\end{aligned} \tag{B.14}$$

The quantities E_ν do not depend on ϕ, and the E_e values do not depend on μ or ϕ. One can calculate the corresponding integrals once [26], using the relations

$$dn_e = \frac{2}{(2\pi\hbar)^3}\int \frac{d\mathbf{p}}{\exp(\frac{\epsilon-\mu_e}{kT})+1}$$

$$\rightarrow \frac{8\pi}{(2\pi\hbar c)^3}\int_{d\epsilon d\mu,\phi=0} \frac{d\epsilon\frac{d\mu}{2}\epsilon^2\beta}{\exp(\frac{\epsilon-\mu_e}{kT})+1}, \tag{B.15}$$

$$1 - F_e(\mathbf{p}) = 1 - \frac{1}{\exp(\frac{\epsilon-\mu_e}{kT})+1} = \frac{\exp(\frac{\epsilon-\mu_e}{kT})}{\exp(\frac{\epsilon-\mu_e}{kT})+1}. \tag{B.16}$$

In the case of anti-neutrino scattering on electrons, one makes the substitution $q \leftrightarrow -q'$ (equivalent to the substitution $C_A \rightarrow -C'_A$) in the matrix element above.

B.2 Absorption of neutrinos by neutrons

The change in the neutrino distribution function due to absorption of neutrinos by neutrons is

$$\left(\frac{\partial f_\nu(\mathbf{q},\mathbf{t})}{\partial t}\right)_{\nu n} = \int \frac{Vd\mathbf{P}Vd\mathbf{Q}Vd\mathbf{p}}{(2\pi\hbar)^6} w_{\mathbf{p},\mathbf{Q};\mathbf{q},\mathbf{P}}\left[-f_\nu(\mathbf{q},t)f_n(\mathbf{P},t)\right],$$

where the probability of this process is

$$w_{\mathbf{p},\mathbf{Q};\mathbf{q},\mathbf{P}} = c(2\pi\hbar)^4\frac{(\hbar c)^4}{V^3}\delta(\mathfrak{q}+\mathfrak{P}-\mathfrak{p}-\mathfrak{Q})\frac{|M_{fi}|^2}{16\epsilon_\nu\epsilon_n\epsilon_e\epsilon_p}, \tag{B.17}$$

and the square of the matrix element $|M_{fi}|^2$ is presented in Refs. [84,300]:

$$|M_{fi}|^2 = 32\left(\frac{G_F}{(\hbar c^3)}c^2\right)^2 \left[(\nu+1)^2(\mathfrak{P}\mathfrak{q})(\mathfrak{Q}\mathfrak{p})\right.$$
$$\left. + (\nu-1)^2(\mathfrak{P}\mathfrak{p})(\mathfrak{Q}\mathfrak{q}) + (\nu^2-1)m_n m_p c^2(\mathfrak{p}\mathfrak{q})\right], \tag{B.18}$$

with $\nu = 1.21$. The conservation of energy–momentum yields the energy of the product particles as a function of their initial momenta and the angle for the final trajectory of the electron. Denote the energy–momentum four vectors $\mathbf{q}$ for ν, $\mathbf{P}$ for n, $\mathbf{p}$ for e, and $\mathbf{Q}$ for p.

The non-standard selection of $\mathbf{Q}$ instead $\mathbf{P}'$ for the protons makes it possible to avoid the use of primes and simplifies our treatment of the reverse reaction.

One should first exclude from the system

$$\epsilon_\nu + \epsilon_n = \epsilon_p + \epsilon_e, \tag{B.19}$$

$$\epsilon_\nu \mathbf{b}_\nu + \beta_n \epsilon_n \mathbf{b}_n = \beta_p \epsilon_p \mathbf{b}_p + \beta_e \epsilon_e \mathbf{b}_e, \tag{B.20}$$

the energy and velocity of p. All terms relating to p on the right-hand sides of the two equations are retained. Taking squares on both sides of the two equations and subtracting the first from the second one yields the equality

$$\begin{aligned}(-m_n^2 - m_e^2 + m_p^2)c^4 - 2\epsilon_\nu \epsilon_n (1 - \beta_n \mathbf{b}_n \mathbf{b}_\nu) + 2\epsilon_e(\epsilon_\nu + \epsilon_n) \\ = 2\epsilon_e \beta_e \mathbf{b}_e (\beta_n \mathbf{b}_n + \epsilon_\nu \mathbf{b}_\nu).\end{aligned} \tag{B.21}$$

Squaring this last equation, one obtains a quadratic equation for the final energy ϵ_e as a function of the angles $\mathbf{b}_e$ and the initial states. The required root satisfies the conservation of energy–momentum.

Further, using a delta function for the momenta, one can integrate the reaction rate over $d\mathbf{Q}$. In the following integration over $d\mathbf{p}$, one should use ϵ_e from the conservation of momentum relation, $\mathbf{p} = \mathbf{q} + \mathbf{P} - \mathbf{Q}$:

$$\begin{aligned}&\int d\epsilon_e \delta(\epsilon_\nu + \epsilon_n - \epsilon_p - \epsilon_e) \\ &\quad = \int d(\epsilon_e + \epsilon_p) \left| \frac{\partial(\epsilon_e + \epsilon_p)}{\partial \epsilon_e} \right|^{-1} \delta(\epsilon_\nu + \epsilon_n - \epsilon_p - \epsilon_e) = |J_{\nu n}|,\end{aligned} \tag{B.22}$$

where

$$J_{\nu n}^{-1} = \frac{\partial(\epsilon_e + \epsilon_p)}{\partial \epsilon_e} = 1 + \frac{\partial(\epsilon_p)}{\partial \epsilon_e}, \tag{B.23}$$

with

$$\beta_p \epsilon_p = (\epsilon_\nu \mathbf{b}_\nu + \beta_n \epsilon_n \mathbf{b}_n - \beta_e \epsilon_e \mathbf{b}_e)\mathbf{b}_p, \tag{B.24}$$

$$\frac{\partial(\beta_p \epsilon_p)}{\partial \epsilon_e} = \frac{1}{\beta_p} \frac{\partial(\epsilon_p)}{\partial \epsilon_e} = -\frac{\partial(\beta_e \epsilon_e)}{\partial \epsilon_e} \mathbf{b}_e \mathbf{b}_p = -\frac{1}{\beta_e} \mathbf{b}_e \mathbf{b}_p. \tag{B.25}$$

The Jacobi matrix transformation is

$$J_{\nu n} = \frac{1}{1 - \beta_p \mathbf{b}_e \mathbf{b}_p / \beta_e}. \tag{B.26}$$

The neutrino absorption coefficient in this reaction is

$$(\chi E)^{\nu n}_{\nu,\omega,k} = \frac{1}{\Delta\epsilon_{\nu,\omega}\Delta\mu_k} \int_{\epsilon_\nu \in \Delta\epsilon_{\nu,\omega}, \mu_\nu \in \Delta\mu_k} dn_\nu dn_n do_e \times J_{\nu n}(1 - F_e(\mathbf{p}))(1 - F_n(\mathbf{Q})) \frac{\epsilon_e^2 \beta_e \epsilon_\nu |M_{fi}|^2 \hbar^2 c^2}{16 \epsilon_\nu \epsilon_n \epsilon_e \epsilon_p}. \tag{B.27}$$

Since the distribution functions for the neutrons, protons, and electrons are independent of angle (in a comoving system), the absorption coefficient likewise does not depend on μ:

$$\chi^{\nu n}_{\nu,\omega,k} = \sum_k \frac{\Delta\mu_k}{2} \chi^{\nu n}_{\nu,\omega,k}. \tag{B.28}$$

This reaction changes the number of electrons and positrons and is included in the equation for $\dot{Y}_e$.

B.3 Creation of neutrinos

The change in the neutrino distribution function due to this process, taking into account the neutron degeneracy, is

$$\left(\frac{\partial f_\nu(\mathbf{q}, \mathbf{t})}{\partial t}\right)_{\nu n} = \int \frac{V d\mathbf{P} V d\mathbf{Q} V d\mathbf{p}}{(2\pi\hbar)^6} w_{\mathbf{q},\mathbf{P};\mathbf{p},\mathbf{Q}} (1 - F_n(\mathbf{P}, t)) f_e(\mathbf{p}, t) f_p(\mathbf{Q}, t). \tag{B.29}$$

The probability $w_{\mathbf{p},\mathbf{Q};\mathbf{q},\mathbf{P}}$ was already determined in the previous section. One can use the conservation of energy–momentum to obtain the energies of the product particles as functions of the initial momenta and the angle of the final trajectory of the neutron. Carrying out the procedure described above yields an equation for ϵ_ν as a function of the angle of the product neutrino $\mathbf{b}_\nu$:

$$\left(m_p^2 + m_e^2 - m_n^2\right) c^4 + 2\epsilon_e\epsilon_p(1 - \beta_e\mathbf{b}_e\beta_p\mathbf{b}_p)$$
$$= 2\epsilon_\nu(\epsilon_p(1 - \beta_p\mathbf{b}_p\mathbf{b}_\nu) + \epsilon_e(1 - \beta_e\mathbf{b}_e\mathbf{b}_\nu)). \quad \text{(B.30)}$$

One can remove the extra solution from the conservation of momentum.

Further, using a delta function, one can integrate the rate over $d\mathbf{P}$. In the following integration over $d\mathbf{q}$, one should use ϵ_p from the previous momentum conservation relation $\mathbf{Q} = \mathbf{P} + \mathbf{q} - \mathbf{p}$, which leads to the Jacobi matrix transformation $J_{ep}^{-1} = 1 - \beta_n\mathbf{b}_\nu\mathbf{b}_n$.

One obtains for the emission coefficient for the neutrino via this reaction

$$\eta_{\nu,\omega,k}^{\nu n} = \frac{1}{\Delta\epsilon_{\nu,\omega}\Delta\mu_k}\int_{\epsilon_\nu\in\Delta\epsilon_{\nu,\omega},\mu_\nu\in\Delta\mu_k} dn_e dn_p do_\nu$$
$$\times |J_{ep}|(1 - F_n(\mathbf{P}, t))\frac{(\epsilon_\nu)^2\epsilon_\nu|M_{fi}|^2\hbar^2c^2}{16\epsilon_e\epsilon_p\epsilon_\nu\epsilon_n}. \quad \text{(B.31)}$$

The absorption coefficient likewise does not depend on μ. This reaction should be taken into account in the equation for $\dot{Y}_e$.

The absorption and emission coefficients for anti-neutrino reaction with nucleons can be considered in the same way. Moreover, the computations include reactions for the absorption and emission of neutrinos by nuclei [171,220].

B.4 Pair creation and annihilation $\nu\tilde{\nu} \rightleftarrows e^-e^+$

The production of neutrinos via the annihilation of electron–positron pairs was also included in the computations. The role of neutrino annihilation in the shell ejection is similar to neutrino scattering on electrons because the number of anti-neutrinos is less compared with neutrinos at the neutralization of the matter (matter is converting in neutrons). The scattering of neutrinos on nucleons (this can be treated analogously to the scattering of neutrinos on electrons) is also included.

The rates of change of the distribution function due to pair creation and annihilation are

$$\left(\frac{\partial f_i(\mathbf{k}_i,t)}{\partial t}\right)_{\nu\tilde{\nu}\to e^-e^+} = -\int d\mathbf{k}_j d\mathbf{p}_- d\mathbf{p}_+ w_{\mathbf{p}_-,\mathbf{p}_+;\mathbf{k}_i,\mathbf{k}_j} \times (1-F_-)(1-F_+) f_i(\mathbf{k}_i,t) f_j(\mathbf{k}_j,t), \tag{B.32}$$

$$\left(\frac{\partial f_i(\mathbf{k}_i,t)}{\partial t}\right)_{\nu\tilde{\nu}\to\nu\tilde{\nu}} = \int d\mathbf{k}_j d\mathbf{p}_- d\mathbf{p}_+ w_{\mathbf{k}_i,\mathbf{k}_j;\mathbf{p}_-,\mathbf{p}_+} f_-(\mathbf{p}_-,t) f_+ \times (\mathbf{p}_+,t), \tag{B.33}$$

for $i=\nu,\ j=\tilde{\nu}$, and for $j=\nu,\ i=\tilde{\nu}$.

$$\left(\frac{\partial f_\pm(\mathbf{p}_\pm,t)}{\partial t}\right)_{\nu\tilde{\nu}\to e^-e^+} = \int d\mathbf{p}_\mp d\mathbf{k}_\nu d\mathbf{k}_{\tilde{\nu}} w_{\mathbf{p}_-,\mathbf{p}_+;\mathbf{k}_\nu,\mathbf{k}_{\tilde{\nu}}} \times (1-F_-)(1-F_+) f_\nu(\mathbf{k}_\nu,t) f_{\tilde{\nu}}(\mathbf{k}_{\tilde{\nu}},t), \tag{B.34}$$

$$\left(\frac{\partial f_\pm(\mathbf{p}_\pm,t)}{\partial t}\right)_{e^-e^+\to\nu\tilde{\nu}} = -\int d\mathbf{p}_\mp d\mathbf{k}_\nu d\mathbf{k}_{\tilde{\nu}} w_{\mathbf{k}_\nu,\mathbf{k}_{\tilde{\nu}};\mathbf{p}_-,\mathbf{p}_+} f_-(\mathbf{p}_-,t) f_+ \times (\mathbf{p}_+,t), \tag{B.35}$$

where

$$w_{\mathbf{p}_-,\mathbf{p}_+;\nu\nu\tilde{\nu},\nu\tilde{\nu}} = \frac{c\delta(\epsilon_- + \epsilon_+ - \epsilon_\nu - \epsilon_{\tilde{\nu}})}{(2\pi\hbar)^2}\delta(\mathbf{p}_- + \mathbf{p}_+ - \mathbf{k}_\nu - \mathbf{k}_{\tilde{\nu}}) \times \frac{|M_{fi}|^2}{16\epsilon_-\epsilon_+\epsilon_\nu\epsilon_{\tilde{\nu}}}. \tag{B.36}$$

Here, the matrix element $|M_{fi}|^2$ is given by Eq. (B.4) by the substitution $\mathfrak{p}' \to -\mathfrak{p}_-$, $\mathfrak{q} \to -\mathfrak{k}$, relabeling the final neutrino and antineutrino momenta by $\mathfrak{k}_\nu \leftrightarrow \mathfrak{k}_{\tilde{\nu}}$, respectively, and multiplying the resultant expression by $1/2$ because the initial spin-sum average is now over an electron and a positron rather than an electron and a neutrino [84,104]:

$$|M_{fi}|^2 = 32\left(\frac{G_F}{(\hbar c^3)}c^2\right)^2 \left[(C_V + C_A)^2(\tilde{\mathfrak{p}}\mathfrak{q})(\mathfrak{p}\tilde{\mathfrak{q}}) \right.$$
$$\left. + (C_V - C_A)^2(\mathfrak{p}\mathfrak{q})(\tilde{\mathfrak{p}}\tilde{\mathfrak{q}}) + (C_V^2 - C_A^2)m_e^2c^2(\mathfrak{q}\tilde{\mathfrak{q}})\right].$$

Calculations of emission and absorption coefficients are similar to annihilation process in Ref. [26]. The energies of neutrinos created via annihilation of a $e^{\pm}$ pair are

$$\epsilon_\nu(\mathbf{b}_\nu) = \frac{m^2c^4 + \epsilon_-\epsilon_+(1 - \beta_-\beta_+\mathbf{b}_-\cdot\mathbf{b}_+)}{\epsilon_-(1 - \beta_-\mathbf{b}_-\cdot\mathbf{b}_\nu) + \epsilon_+(1 - \beta_+\mathbf{b}_+\cdot\mathbf{b}_\nu)},$$
$$\epsilon_{\tilde{\nu}}(\mathbf{b}_\nu) = \epsilon_- + \epsilon_+ - \epsilon_\nu, \tag{B.37}$$

while the energies of pair particles created by two photons are

$$\epsilon_-(\mathbf{b}_-) = \frac{B \mp \sqrt{B^2 - AC}}{A}, \qquad \epsilon_+(\mathbf{b}_-) = \epsilon_\nu + \epsilon_{\tilde{\nu}} - \epsilon_-, \tag{B.38}$$

where $A = (\epsilon_\nu + \epsilon_{\tilde{\nu}})^2 - [(\epsilon_\nu\mathbf{b}_\nu + \epsilon_{\tilde{\nu}}\mathbf{b}_{\tilde{\nu}})\cdot\mathbf{b}_-]^2$, $B = (\epsilon_\nu + \epsilon_{\tilde{\nu}})\epsilon_\nu\epsilon_{\tilde{\nu}}(1 - \mathbf{b}_\nu\cdot\mathbf{b}_{\tilde{\nu}})$, and $C = m_e^2c^4[(\epsilon_\nu\mathbf{b}_\nu + \epsilon_{\tilde{\nu}}\mathbf{b}_{\tilde{\nu}})\cdot\mathbf{b}_-]^2 + \epsilon_\nu^2\epsilon_{\tilde{\nu}}^2(1 - \mathbf{b}_\nu\cdot\mathbf{b}_{\tilde{\nu}})^2$. Only one root in Eq. (B.38) has to be chosen. From energy–momentum conservation

$$\mathfrak{k}_\nu + \mathfrak{k}_{\tilde{\nu}} - \mathfrak{p}_- = \mathfrak{p}_+, \tag{B.39}$$

we have, taking square from the energy part,

$$\epsilon_\nu^2 + \epsilon_{\tilde{\nu}}^2 + \epsilon_-^2 + 2\epsilon_\nu\epsilon_{\tilde{\nu}} - 2\epsilon_\nu\epsilon_- - 2\epsilon_{\tilde{\nu}}\epsilon_- = \epsilon_+^2. \tag{B.40}$$

Taking square from the momentum part

$$\epsilon_\nu^2 + \epsilon_{\tilde{\nu}}^2 + \epsilon_-^2\beta_-^2 + 2\epsilon_\nu\epsilon_{\tilde{\nu}}\mathbf{b}_\nu\cdot\mathbf{b}_{\tilde{\nu}} - 2\epsilon_\nu\epsilon_-\beta_-\mathbf{b}_\nu\cdot\mathbf{b}_- - 2\epsilon_{\tilde{\nu}}\epsilon_-\beta_-\mathbf{b}_{\tilde{\nu}}\cdot\mathbf{b}_- = (\epsilon_+\beta_+)^2, \tag{B.41}$$

there are no additional roots because of the arbitrary $\mathbf{e}_+$, or

$$\epsilon_\nu\epsilon_{\tilde{\nu}}(1 - \mathbf{b}_\nu\cdot\mathbf{b}_{\tilde{\nu}}) - \epsilon_\nu\epsilon_-(1 - \beta_-\mathbf{b}_\nu\cdot\mathbf{b}_-) - \epsilon_{\tilde{\nu}}\epsilon_-(1 - \beta\mathbf{b}_{\tilde{\nu}}\cdot\mathbf{b}_-) = 0,$$
$$\epsilon_-\beta_-(\epsilon_\nu\mathbf{b}_\nu + \epsilon_{\tilde{\nu}}\mathbf{b}_{\tilde{\nu}})\cdot\mathbf{b}_- = \epsilon_-(\epsilon_\nu + \epsilon_{\tilde{\nu}}) - \epsilon_\nu\epsilon_{\tilde{\nu}}(1 - \mathbf{b}_\nu\cdot\mathbf{b}_{\tilde{\nu}}). \tag{B.42}$$

Getting rid of β, we find

$$\begin{aligned}\epsilon_\nu^2\epsilon_{\tilde{\nu}}^2(1-\mathbf{b}_\nu\cdot\mathbf{b}_{\tilde{\nu}})^2 - 2\epsilon_\nu\epsilon_{\tilde{\nu}}(1-\mathbf{b}_\nu\cdot\mathbf{b}_{\tilde{\nu}})(\epsilon_\nu+\epsilon_{\tilde{\nu}})\epsilon_-\\ +\left\{(\epsilon_\nu+\epsilon_{\tilde{\nu}})^2 - [(\epsilon_\nu\mathbf{b}_\nu+\epsilon_{\tilde{\nu}}\mathbf{b}_{\tilde{\nu}})\cdot\mathbf{b}_-]^2\right\}\epsilon_-^2\\ = [(\epsilon_\nu\mathbf{b}_\nu+\epsilon_{\tilde{\nu}}\mathbf{b}_{\tilde{\nu}})\cdot\mathbf{b}_-](-m^2).\end{aligned} \tag{B.43}$$

Therefore, the condition to be checked reads

$$\begin{aligned}&\epsilon_-\beta_-[(\epsilon_\nu\mathbf{b}_\nu+\epsilon_{\tilde{\nu}}\mathbf{b}_{\tilde{\nu}})\cdot\mathbf{b}_-]^2\\ &\quad = [\epsilon_-(\epsilon_\nu+\epsilon_{\tilde{\nu}}) - (\epsilon_\nu\epsilon_{\tilde{\nu}})(1-\mathbf{b}_\nu\cdot\mathbf{b}_{\tilde{\nu}})][(\epsilon_\nu\mathbf{b}_\nu+\epsilon_{\tilde{\nu}}\mathbf{b}_{\tilde{\nu}})\cdot\mathbf{b}_-] \geq 0.\end{aligned} \tag{B.44}$$

Finally, integration of Eqs. (B.32)–(B.35) yields

$$\eta_{\nu,\omega,k}^{e^-e^+\to\nu\tilde{\nu}} = \frac{1}{\Delta\epsilon_{\nu,\omega}\Delta\mu_k}\int_{\substack{\epsilon_\nu\in\Delta\epsilon_{\gamma\nu,\omega},\\ \mu_\nu\in\Delta\mu_k}} dn_{e^-}dn_{e^+}do_\nu J_{\rm ca}\frac{\epsilon_\nu^2|M_{fi}|^2}{16\epsilon_-\epsilon_+\epsilon_{\tilde{\nu}}}, \tag{B.45}$$

$$\eta_{\tilde{\nu},\omega,k}^{e^-e^+\to\nu\tilde{\nu}} = \frac{1}{\Delta\epsilon_{\nu,\omega}\Delta\mu_k}\int_{\substack{\epsilon_{\tilde{\nu}}\in\Delta\epsilon_{\nu,\omega},\\ \mu_{\tilde{\nu}}\in\Delta\mu_k}} dn_{e^-}dn_{e^+}do_\nu J_{\rm ca}\frac{\epsilon_\nu|M_{fi}|^2}{16\epsilon_-\epsilon_+}, \tag{B.46}$$

$$(\chi E)_{e^-,\omega,k}^{e^-e^+\to\nu\tilde{\nu}} = \frac{1}{\Delta\epsilon_{e,\omega}\Delta\mu_k}\int_{\substack{\epsilon_-\in\Delta\epsilon_{e,\omega},\\ \mu_-\in\Delta\mu_k}} dn_{e^-}dn_{e^+}do_\nu J_{\rm ca}\frac{\epsilon_\nu|M_{fi}|^2}{16\epsilon_+\epsilon_{\tilde{\nu}}}, \tag{B.47}$$

$$(\chi E)_{e^+,\omega,k}^{e^-e^+\to\nu\tilde{\nu}} = \frac{1}{\Delta\epsilon_{e,\omega}\Delta\mu_k} + \int_{\substack{\epsilon_+\in\Delta\epsilon_{e,\omega},\\ \mu_+\in\Delta\mu_k}} dn_{e^-}dn_{e^+}do_\nu J_{\rm ca}\frac{\epsilon_\nu|\tilde{M}_{fi}|^2}{16\epsilon_-\epsilon_{\tilde{\nu}}}, \tag{B.48}$$

$$\begin{aligned}(\chi E)_{\nu,\omega,k}^{\nu\tilde{\nu}\to e^-e^+} &= \frac{1}{\Delta\epsilon_{\nu,\omega}\Delta\mu_k}\int_{\substack{\epsilon_\nu\in\Delta\epsilon_{\nu,\omega},\\ \mu_\nu\in\Delta\mu_k}} (1-F_-)(1-F_+)dn_\nu dn_{\tilde{\nu}}do_- J_{\rm ca}\\ &\quad\times\frac{\epsilon_-\beta_-|M_{fi}|^2}{16\epsilon_{\tilde{\nu}}\epsilon_+},\end{aligned} \tag{B.49}$$

$$(\chi E)^{\nu\tilde{\nu}\to e^-e^+}_{\tilde{\nu},\omega,k} = \frac{1}{\Delta\epsilon_{\nu,\omega}\Delta\mu_k}\int_{\substack{\epsilon_{\tilde{\nu}}\in\Delta\epsilon_{\nu,\omega},\\ \mu_{\tilde{\nu}}\in\Delta\mu_k}} (1-F_-)(1-F_+)dn_\nu dn_{\tilde{\nu}}do_- J_{\rm ca} \times \frac{\epsilon_-\beta_-|M_{fi}|^2}{16\epsilon_\nu\epsilon_+}, \tag{B.50}$$

$$\eta^{\nu\tilde{\nu}\to e^-e^+}_{e^-,\omega,k} = \frac{1}{\Delta\epsilon_{e,\omega}\Delta\mu_k}\int_{\substack{\epsilon_-\in\Delta\epsilon_{e,\omega},\\ \mu_-\in\Delta\mu_k}} (1-F_-)(1-F_+)dn_\nu dn_{\tilde{\nu}}do_- J_{\rm ca} \times \frac{\epsilon_-^2\beta_-|M_{fi}|^2}{16\epsilon_\nu\epsilon_{\tilde{\nu}}\epsilon_+}, \tag{B.51}$$

$$\eta^{\nu\tilde{\nu}\to e^-e^+}_{e^+,\omega,k} = \frac{1}{\Delta\epsilon_{e,\omega}\Delta\mu_k}\int_{\substack{\epsilon_+\in\Delta\epsilon_{e,\omega},\\ \mu_+\in\Delta\mu_k}} (1-F_-)(1-F_+)dn_\nu dn_{\tilde{\nu}}do_- J_{\rm ca} \times \frac{\epsilon_-\beta_-|M_{fi}|^2}{16\epsilon_\nu\epsilon_{\tilde{\nu}}}, \tag{B.52}$$

where $dn_\pm = d\epsilon_\pm do_\pm\epsilon_\pm^2\beta_\pm f_\pm$, $dn_{\nu,\tilde{\nu}} = d\epsilon_{\nu,\tilde{\nu}}do_{\nu,\tilde{\nu}}\epsilon^2_{\nu,\tilde{\nu}}f_{\nu,\tilde{\nu}}$ and

$$J_{\rm ca} = \frac{\epsilon_+\beta_-}{(\epsilon_+ + \epsilon_-)\,\beta_- - (\epsilon_\nu\mathbf{b}_\nu + \epsilon_{\tilde{\nu}}\mathbf{b}_{\tilde{\nu}})\cdot\mathbf{b}_-}. \tag{B.53}$$

The quantities E_ν do not depend on ϕ, and the E_e values do not depend on μ or ϕ. One can calculate the corresponding integrals once [26], using the relations

$$dn_\mp = \frac{8\pi}{(2\pi\hbar c)^3}\int_{d\epsilon d\mu,\phi=0}\frac{d\epsilon_\mp\frac{d\mu}{2}\epsilon_\mp^2\beta_m p}{\exp(\frac{\epsilon_\mp-\mu_\mp}{kT})+1}, \tag{B.54}$$

$$1 - F_\mp(\mathbf{p}_\mp) = 1 - \frac{1}{\exp(\frac{\epsilon_\mp-\mu_\mp}{kT})+1}. \tag{B.55}$$

Isotropic distribution of neutrino is assumed in the calculations of its annihilation $\nu + \tilde{\nu} \rightleftarrows e + e^+$ (in exact calculations, the rate of the inverse reaction depends on the two angles for the two neutrinos).

B.5 Neutrino scattering on nucleons $\nu + N \to \nu' + N'$

Matrix element is [84]

$$|M_{fi}|^2 = 32\left(\frac{G_F}{(\hbar c^3)}c^2\right)^2 \left[(h_V^N + h_A^N)^2(\tilde{\mathfrak{p}}\mathfrak{q})(\mathfrak{p}\tilde{\mathfrak{q}})\right.$$

$$\left. + (H_V^N - h_A^N)^2(\mathfrak{p}\mathfrak{q})(\tilde{\mathfrak{p}}\tilde{\mathfrak{q}}) + ((h_V^N)^2 - (h_A^N)^2)m_e^2c^2(\mathfrak{q}\tilde{\mathfrak{q}})\right],$$

$h_V^p = 0.5 - 2\sin^2\theta_W$, $h_A^p = 0.5g_a$, and $h_V^n = -0.5$, $h_A^n = -0.5g_a$. Calculations of emission and absorption coefficients are similar to neutrino scattering on electrons $\nu + e$.

References

[1] J. Abadie, *et al.* Search for gravitational waves from compact binary coalescence in LIGO and Virgo data from S5 and VSR1. *Physical Review D*, 82(10):102001, November 2010.

[2] B. P. Abbott, *et al.* Search for gravitational waves from low mass compact binary coalescence in 186 days of LIGO's fifth science run. *Physical Review D*, 80(4):047101, August 2009.

[3] K. Abe, *et al.* Supernova model discrimination with hyper-kamiokande. *The Astrophysical Journal*, 916(1):15, July 2021.

[4] N. Y. Agafonova, *et al.* Implication for the Core-collapse supernova rate from 21 years of data of the large volume detector. *The Astrophysical Journal*, 802:47, March 2015.

[5] E. N. Akimova. Parallel Gauss algorithms for block tridiagonal linear systems. *Matematicheskoe Modelirovanie*, 6(9):61–67, 1994.

[6] A. G. Aksenov. A study of dynamical stability of a rotating n=1.5 polytrope. *Astronomy Letters*, 22(5):634–647, September 1996.

[7] A. G. Aksenov. Statement of the problem of gravitational collapse of a stellar iron-oxygen core and a numerical method of its solution. *Astronomy Letters*, 24:482–490, July 1998.

[8] A. G. Aksenov. Computing the collapse of iron-oxygen stellar cores with allowance for the absorption and emission of electron neutrinos and antineutrinos. *Astronomy Letters*, 25:307–317, May 1999.

[9] A. G. Aksenov. Gravitational radiation during fragmentation of a rotating neutron star. *Astronomy Letters*, 25(3):127–135, March 1999.

[10] A. G. Aksenov. Numerical solution of the Poisson equation for the three-dimensional modeling of stellar evolution. *Astronomy Letters*, 25(3):185–190, March 1999.

[11] A. G. Aksenov. Computation of shock waves in plasma. *Computational Mathematics and Mathematical Physics*, 55:1752–1769, October 2015.

[12] A. G. Aksenov and S. I. Blinnikov. A Newton iteration method for obtaining equilibria of rapidly rotating stars. *Astronomy and Astrophysics*, 290, October 1994.

[13] A. G. Aksenov, S. I. Blinnikov, and V. S. Imshennik. Rapidly rotating cool neutron stars. *Astronomicheskij Zhurnal*, 72:717, October 1995.

[14] A. G. Aksenov and V. M. Chechetkin. Computations of the collapse of a stellar iron core allowing for the absorption, emission, and scattering of electron neutrinos and anti-neutrinos. *Astronomy Reports*, 56:193–206, March 2012.

[15] A. G. Aksenov and V. M. Chechetkin. Gravitational radiation during coalescence of neutron stars. *Astronomy Reports*, 57(7):498–508, July 2013.

[16] A. G. Aksenov and V. M. Chechetkin. Supernova explosion mechanism taking into account large-scale convection and neutrino transport. *Astronomy Reports*, 58:442–450, July 2014.

[17] A. G. Aksenov and V. M. Chechetkin. Neutronization of matter in a Stellar core and convection during gravitational collapse. *Astronomy Reports*, 60:655–668, July 2016.

[18] A. G. Aksenov and V. M. Chechetkin. Large-scale instability during gravitational collapse with neutrino transport and a core-collapse supernova. *Astronomy Reports*, 62:251–263, April 2018.

[19] A. G. Aksenov and V. M. Chechetkin. Large-scale instability during gravitational collapse and the escaping neutrino spectrum during a supernova explosion. *Astronomy Reports*, 63(11):900–909, November 2019.

[20] A. G. Aksenov and V. M. Chechetkin. Large-scale instability in supernovae and the neutrino spectrum. *Astronomy Reports*, 65(10):916–920, October 2021.

[21] A. G. Aksenov and V. M. Chechetkin. Nonequilibrium neutronization and large-scale convection in gravitational collapse. *Astronomy Reports*, 66(1):1–11, January 2022.

[22] A. G. Aksenov and V. M. Chechetkin. Large-scale convection during gravitational collapse with neutrino transport in 2D and 3D models on fine grids. *Astronomy Reports*, 67(3):209–219, March 2023.

[23] A. G. Aksenov and M. D. Churazov. Deuterium targets and the MDMT code. *Laser and Particle Beams*, 21:81–84, January 2003.

[24] A. G. Aksenov and V. S. Imshennik. A numerical study of the stability of a rapidly rotating neutron star: Axially symmetric model. *Astronomy Letters*, 20(1):24–39, January 1994.

[25] A. G. Aksenov, M. Milgrom, and V. V. Usov. Radiation from hot, bare, strange stars. *Monthly Notices of the Royal Astronomical Society*, 343:L69–L72, August 2003.

[26] A. G. Aksenov, M. Milgrom, and V. V. Usov. Structure of pair winds from compact objects with application to emission from hot bare strange stars. *The Astrophysical Journal*, 609:363–377, July 2004.

[27] A. G. Aksenov, M. Milgrom, and V. V. Usov. Pair winds in schwarzschild spacetime with application to hot bare strange stars. *The Astrophysical Journal*, 632:567–575, October 2005.

[28] A. G. Aksenov and D. K. Nadyozhin. The calculation of a collapse of iron-oxygen stellar core in one-group energy approximation. *Astronomy Letters*, 24:703–715, November 1998.

[29] A. G. Aksenov, V. F. Tishkin, and V. M. Chechetkin. Godunov-type method and Shafranov's task for multi-temperature plasma. *Mathematical Models and Computer Simulations*, 11:360–373, 2019.

[30] A. G. Aksenov. A Godunov-type method for a multi-temperature plasma with the tabulated equation of state. In *Materials Science and Engineering Conference Series*, volume 927. *XIII International Conference on Applied Mathematics and Mechanics in the Aerospace Industry (AMMAI'2020)*, p. 012037. Alushta, Russia, 6–13 September 2020.

[31] C. Alcock, E. Farhi, and A. Olinto. Strange stars. *The Astrophysical Journal*, 310:261–272, November 1986.

[32] E. N. Alekseev, *et al.* Possible detection of a neutrino signal on 23 February 1987 at the Baksan underground scintillation telescope of the institute of nuclear research. *Soviet Journal of Experimental and Theoretical Physics Letters*, 45:589, May 1987.

[33] M. L. Alme and J. R. Wilson. X-ray emission from a neutron star accreting material. *The Astrophysical Journal*, 186:1015–1026, December 1973.

[34] S. I. Anisimov, *et al.* Ablated matter expansion and crater formation under the action of ultrashort laser pulse. *Soviet Journal of Experimental and Theoretical Physics*, 103:183–197, August 2006.

[35] N. V. Ardeljan, G. S. Bisnovatyi-Kogan, and S. G. Moiseenko. Magnetorotational supernovae. *Monthly Notices of the Royal Astronomical Society*, 359:333–344, May 2005.

[36] D. Arnett. Mass dependence in gravitational collapse of stellar cores. *Canadian Journal of Physics*, 45:1621–1641, 1967.

[37] W. D. Arnett. A possible model of supernovae: Detonation of ^{12}C. *Astrophysics and Space Science*, 5:180–212, October 1969.

[38] W. D. Arnett. Supernova theory and supernova 1987A. *The Astrophysical Journal*, 319:136–142, August 1987.

[39] R. Arnowitt and S. Deser. Quantum theory of gravitation: General formulation and linearized theory. *Physical Review*, 113(2):745–750, January 1959.

[40] R. Arnowitt, S. Deser, and C. W. Misner. Dynamical structure and definition of energy in general relativity. *Physical Review*, 116(5):1322–1330, December 1959.

[41] R. Arnowitt, S. Deser, and C. W. Misner. Energy and the criteria for radiation in general relativity. *Physical Review*, 118(4):1100–1104, May 1960.

[42] R. Arnowitt, S. Deser, and C. W. Misner. Finite self-energy of classical point particles. *Physical Review Letters*, 4(7):375–377, April 1960.

[43] R. Arnowitt, S. Deser, and C. W. Misner. Gravitational-electromagnetic coupling and the classical self-energy problem. *Physical Review*, 120(1):313–320, October 1960.

[44] R. Arnowitt, S. Deser, and C. W. Misner. Interior schwarzschild solutions and interpretation of source terms. *Physical Review*, 120(1):321–324, October 1960.

[45] R. Arnowitt, S. Deser, and C. W. Misner. Coordinate invariance and energy expressions in general relativity. *Physical Review*, 122(3):997–1006, May 1961.

[46] R. Arnowitt, S. Deser, and C. W. Misner. Wave zone in general relativity. *Physical Review*, 121(5):1556–1566, March 1961.

[47] R. Arnowitt, S. Deser, and C. W. Misner. Republication of: The dynamics of general relativity. *General Relativity and Gravitation*, 40(9):1997–2027, September 2008.

[48] K. G. Arun, *et al.* Publisher's note: Higher signal harmonics, LISA's angular resolution, and dark energy. *Physical Review D*, 76(12):129903, December 2007.

[49] B. Aylott, *et al.* Testing gravitational-wave searches with numerical relativity waveforms: Results from the first Numerical INJection Analysis (NINJA) project. *Classical and Quantum Gravity*, 26(16):165008, August 2009.

[50] W. Baade and F. Zwicky. Cosmic rays from super-novae. *Proceedings of the National Academy of Science*, 20:259–263, May 1934.

[51] W. Baade and F. Zwicky. On super-novae. *Proceedings of the National Academy of Science*, 20:254–259, May 1934.

[52] I. V. Baikov and V. M. Chechetkin. The influence of high-energy neutrinos on the ejection of the envelope of a type-II supernova. *Astronomy Reports*, 48:229–235, March 2004.

[53] I. V. Baikov, *et al.* Radiation of a neutrino mechanism for type II supernovae. *Astronomy Reports*, 51:274–281, April 2007.

[54] A. A. Baranov and V. M. Chechetkin. Did the SN 1987A outburst leave a compact remnant? *Astronomy Reports*, 55:525–531, June 2011.

[55] Z. Barkat, *et al.* On the collapse of iron stellar cores. *The Astrophysical Journal*, 196:633–638, March 1975.

[56] M. M. Basko. Stopping of fast ions in a dense plasma. *Soviet Journal of Plasma Physics*, 10:689–694, 1984.

[57] M. M. Basko, M. D. Churazov, and A. G. Aksenov. Prospects of heavy ion fusion in cylindrical geometry. *Laser and Particle Beams*, 20:411–414, July 2002.

[58] O. M. Belotserkovskii, A. M. Oparin, and V. M. Chechetkin. *Turbulence: New Approaches.* Cambridge International Science Publishing, 2005.

[59] V. B. Berestetskii, E. M. Lifshitz, and V. B. Pitaevskii. *Quantum Electrodynamics.* Elsevier, 1982.

[60] V. S. Berezinskii, *et al.* On the possibility of a two-bang supernova collapse. *Nuovo Cimento C Geophysics Space Physics C*, 11:287–303, June 1988.

[61] H. Bethe and R. Peierls. The neutrino. *Nature*, 133:532, April 1934.

[62] H. A. Bethe. Supernova mechanisms. *Reviews of Modern Physics*, 62:801–866, October 1990.

[63] H. A. Bethe and J. R. Wilson. Revival of a stalled supernova shock by neutrino heating. *The Astrophysical Journal*, 295:14–23, August 1985.

[64] R. M. Bionta, *et al.* Observation of a neutrino burst in coincidence with supernova 1987A in the Large Magellanic Cloud. *Physical Review Letters*, 58:1494–1496, April 1987.

[65] G. S. Bisnovatyi-Kogan. The explosion of a rotating star as a supernova mechanism. *Astronomicheskij Zhurnal*, 47:813, August 1970.

[66] G. S. Bisnovatyi-Kogan. The explosion of a rotating star as a supernova mechanism. *Soviet Astronomy*, 14:652, February 1971.

[67] G. S. Bisnovatyi-Kogan. A self-consistent solution for an accretion disc structure around a rapidly rotating non-magnetized star. *Astronomy and Astrophysics*, 274:796, July 1993.

[68] G. S. Bisnovatyi-Kogan. Analytic solution for kinetic equilibrium with respect to beta-processes in nucleon plasmas with relativistic pairs. *Astrophysics*, 55:387–396, September 2012.

[69] G. S. Bisnovatyi-Kogan and S. I. Blinnikov. Static criteria for stability of arbitrarily rotating stars. *Astronomy and Astrophysics*, 31:391, April 1974.

[70] G. S. Bisnovatyi-Kogan and S. I. Blinnikov. The equilibrium of a rotating gaseous disk of finite thickness. *Soviet Astronomy*, 25:175, April 1981.

[71] G. S. Bisnovatyj-Kogan. *Physical Problems of the Theory of Stellar Evolution.* Moskva (USSR): Nauka, 1989.

[72] L. Blanchet, T. Damour, and G. Schaefer. Post-Newtonian hydrodynamics and post-Newtonian gravitational wave generation for numerical relativity. *Monthly Notices of the Royal Astronomical Society*, 242:289–305, January 1990.

[73] J. Blazek. *Computational Fluid Dynamics Principles and Applications*. Elsevier, 2001.

[74] S. I. Blinnikov. Self-consistent field method in the theory of rotating stars. *Astronomicheskij Zhurnal*, 52:243, April 1975.

[75] S. I. Blinnikov, N. V. Dunina-Barkovskaya, and D. K. Nadyozhin. Equation of state of a Fermi gas: Approximations for various degrees of relativism and degeneracy. *The Astrophysical Journal Supplement Series*, 106:171, September 1996.

[76] S. I. Blinnikov and A. M. Khokhlov. Development of detonations in degenerate stars. *Soviet Astronomy Letters*, 12:131–134, April 1986.

[77] S. I. Blinnikov, P. V. Sasorov, and S. E. Woosley. Self-acceleration of nuclear flames in supernovae. *Space Science Reviews*, 74(3–4):299–311, November 1995.

[78] J. M. Blondin, A. Mezzacappa, and C. DeMarino. Stability of standing accretion shocks, with an eye toward core-collapse supernovae. *The Astrophysical Journal*, 584:971–980, February 2003.

[79] P. Bodenheimer. Rapidly rotating stars. VII. Effects of angular momentum on upper-main models. *The Astrophysical Journal*, 167:153, July 1971.

[80] P. Bodenheimer and Jeremiah P. Ostriker. Rapidly rotating stars. VIII. Zero-viscosity polytropic sequences. *The Astrophysical Journal*, 180:159–170, February 1973.

[81] A. R. Bodmer. Collapsed nuclei. *Physical Review D*, 4:1601–1606, September 1971.

[82] L. Boltzmann. *Lectures on Gas Theory*. Dover Publications, 2011.

[83] S. W. Bruenn. The effect of beta processes on the dynamic evolution of carbon-detonation supernovae. *The Astrophysical Journal*, 168:203, September 1971.

[84] S. W. Bruenn. Stellar core collapse: Numerical model and infall epoch. *The Astrophysical Journal Supplement Series*, 58:771–841, August 1985.

[85] S. W. Bruenn. Neutrinos from SN1987A and current models of stellar-core collapse. *Physical Review Letters*, 59:938–941, August 1987.

[86] A. Burrows and D. Vartanyan. Core-collapse supernova explosion theory. *Nature*, 589(7840):29–39, January 2021.

[87] A. Burrows. Convection and the mechanism of type II supernovae. *The Astrophysical Journal*, 318:L57, July 1987.

[88] V. Bychkov, *et al.* Dynamics of bubbles in supernovae and turbulent vortices. *Astronomy Reports*, 50:298–311, April 2006.

[89] V. V. Bychkov and Michael A. Liberman. Thermal instability and pulsations of the flame front in white dwarfs. *The Astrophysical Journal*, 451:711, October 1995.

[90] J.-F. Caron and A. R. Zhitnitsky. Bremsstrahlung emission from quark stars. *Physical Review D*, 80(12):123006, December 2009.

[91] J. I. Castor. Radiative transfer in spherically symmetric flows. *The Astrophysical Journal*, 178:779–792, December 1972.

[92] G. R. Caughlan and William A. Fowler. Thermonuclear reaction rates V. *Atomic Data and Nuclear Data Tables*, 40:283, January 1988.

[93] D. A. Chamulak, *et al.* Asymmetry and the nucleosynthetic signature of nearly edge-lit detonation in white dwarf cores. *The Astrophysical Journal*, 744(1):27, January 2012.

[94] S. Chandrasekhar. *Ellipsoidal Figures of Equilibrium*. Yale University Press, 1969.

[95] S. Chandrasekhar and N. R. Lebovitz. *The Astrophysical Journal*, 138:185, July 1963.

[96] P. Chardonnet, V. Chechetkin, and L. Titarchuk. On the pair-instability supernovae and gamma-ray burst phenomenon. *Astrophysics and Space Science*, 325:153–161, February 2010.

[97] J. G. Charney, R. Fjörtoft, and J. von Neumann. Numerical Integration of the Barotropic Vorticity Equation. *Tellus*, 2:237, August 1950.

[98] V. M. Chechetkin. Neutronization of matter at high densities. *Astronomicheskij Zhurnal*, 46:206, January 1969.

[99] V. M. Chechetkin and A. G. Aksenov. Supernova-explosion mechanism involving neutrinos. *Physics of Atomic Nuclei*, 81:128–138, 2018.

[100] V. M. Chechetkin, *et al.* Types I and II supernovae and the neutrino mechanism of thermonuclear explosion of degenerate carbon-oxygen stellar cores. *Astrophysics and Space Science*, 67:61–97, January 1980.

[101] V. M. Chechetkin, *et al.* On the neutrino mechanism of supernova explosions. *Astronomy Letters*, 23(1):30–36, January 1997.

[102] A. Chen, T. Yu, and R. Xu. The birth of quark stars: Photon-driven supernovae? *The Astrophysical Journal*, 668:L55–L58, October 2007.

[103] R. A. Chevalier. Self-similar solutions for the interaction of stellar ejecta with an external medium. *The Astrophysical Journal*, 258:790–797, July 1982.

[104] H.-Y. Chiu and R. C. Stabler. Emission of photoneutrinos and pair annihilation neutrinos from stars. *Physical Review*, 122:1317–1322, May 1961.

[105] A. Chodos, *et al.* New extended model of hadrons. *Physical Review D*, 9:3471–3495, June 1974.

[106] M. J. Clement. On the solution of Poisson's equation for rapidly rotating stars. *The Astrophysical Journal*, 194:709–714, December 1974.

[107] P. Colella and H. M. Glaz. Efficient solution algorithms for the Riemann problem for real gases. *Journal of Computational Physics*, 59:264–289, June 1985.

[108] P. Colella and P. R. Woodward. The piecewise parabolic method (PPM) for gas-dynamical simulations. *Journal of Computational Physics*, 54:174–201, September 1984.

[109] S. A. Colgate and R. H. White. The hydrodynamic behavior of supernovae explosions. *The Astrophysical Journal*, 143:626, March 1966.

[110] M. Colpi, S. L. Shapiro, and S. A. Teukolsky. A hydrodynamical model for the explosion of a neutron star just below the minimum mass. *The Astrophysical Journal*, 414:717, September 1993.

[111] S. M. Couch and C. D. Ott. Revival of the stalled core-collapse supernova shock triggered by precollapse asphericity in the progenitor star. *The Astrophysical Journal*, 778:L7, November 2013.

[112] S. M. Couch and C. D. Ott. The role of turbulence in neutrino-driven core-collapse supernova explosions. *The Astrophysical Journal*, 799:5, January 2015.

[113] F. J. Courant, R. *Introduction to Calculus and Analysis I.* Springer Science + Business Media, New York, 1989.

[114] R. Courant, K. Friedrichs, and H. Lewy. Über die partiellen Differenzengleichungen der mathematischen Physik. *Mathematische Annalen*, 100:32–74, 1928.

[115] R. Courant and D. Hilbert. *Methods of Mathematical Physics.* 1989.

[116] J. Crank, P. Nicolson, and D. R. Hartree. A practical method for numerical evaluation of solutions of partial differential equations of the heat-conduction type. *Proceedings of the Cambridge Philosophical Society*, 43:50, 1947.

[117] Z. G. Dai and K. S. Cheng. Properties of dense matter in a supernova core in the relativistic mean-field theory. *Astronomy and Astrophysics*, 330:569–577, February 1998.

[118] S. R. de Groot, W. A. van Leeuwen, and Ch. G. Weert. *Relativistic Kinetic Theory: Principles and Applications.* North Holland, 1980.

[119] J. Deng, *et al.* On the light curve and spectrum of SN 2003dh separated from the optical afterglow of GRB 030329. *The Astrophysical Journal*, 624(2):898–905, May 2005.

[120] L. Dessart, *et al.* The proto-neutron star phase of the collapsar model and the route to long-soft gamma-ray bursts and hypernovae. *The Astrophysical Journal*, 673:L43, January 2008.

[121] J. C. Dolence, A. Burrows, and W. Zhang. Two-dimensional core-collapse supernova models with multi-dimensional transport. *The Astrophysical Journal*, 800:10, February 2015.

[122] J. J. Duderstadt and G. A. Moses. *Inertial Confinement Fusion*. John Wiley & Sons, New York, 1982.

[123] Y. Eriguchi and E. Mueller. A general computational method for obtaining equilibria of self-gravitating and rotating gases. *Astronomy and Astrophysics*, 146(2):260–268, May 1985.

[124] Y. Eriguchi and E. Mueller. Equilibrium models of differentially rotating polytropes and the collapse of rotating stellar cores. *Astronomy and Astrophysics*, 147(1):161–168, June 1985.

[125] Y. Eriguchi and E. Mueller. Structure of rapidly rotating axisymmetric stars. I: A numerical method for stellar structure and meridional circulation. *Astronomy and Astrophysics*, 248(2): 435–447, August 1991.

[126] Y. Eriguchi and E. Mueller. Structure and circulation of self-gravitating toroids. *The Astrophysical Journal*, 416:666, October 1993.

[127] J. J. Erpenbeck. Stability of idealized one-reaction detonations. *Physics of Fluids*, 7(5):684–696, May 1964.

[128] B. D. Farris, Y. T. Liu, and S. L. Shapiro. Binary black hole mergers in gaseous disks: Simulations in general relativity. *Physical Review D*, 84(2):024024, July 2011.

[129] M. Feroci, *et al.* The giant flare of 1998 August 27 from SGR 1900+14. I. An interpretive study of BeppoSAX and Ulysses observations. *The Astrophysical Journal*, 549:1021–1038, March 2001.

[130] C. A. J. Fletcher (editor). *Computational Techniques for Fluid Dynamics. Volume 1: Fundamental and General Techniques. Volume 2: Specific Techniques for Different Flow Categories*, volume 1, 1988.

[131] V. E. Fortov, D. H. Hoffmann, and B. Y. Sharkov. Intense ion beams for generating extreme states of matter. *Physics Uspekhi*, 51:109–131, February 2008.

[132] W. A. Fowler and F. Hoyle. Neutrino processes and pair formation in massive stars and supernovae. *The Astrophysical Journal Supplement Series*, 9:201, December 1964.

[133] K. Fricke. Instabilität stationärer rotation in sternen. *Zeitschrift fr Astrophysik*, 68:317, January 1968.

[134] S. A. Galkin, *et al.* A finite-difference adaptive grid method for computing the equilibria of rotating self-gravitating barotropic gases. *Astronomy and Astrophysics*, 269(1–2):255–266, March 1993.

[135] G. Gamow and M. Schoenberg. The possible role of neutrinos in stellar evolution. *Physical Review*, 58:1117, December 1940.

[136] G. Gamow and M. Schoenberg. Neutrino theory of stellar collapse. *Physical Review*, 59:539–547, April 1941.

[137] C. W. Gear. *Numerical Initial Value Problems in Ordinary Differential Equations.* Prentice-Hall, Inc. Englewood Cliffs, NJ, 1971.

[138] S. S. Gershtein, *et al.* Neutrino mechanism of thermonuclear burning of carbon, formation of neutrino stars, and supernova outbursts. *Soviet Journal of Experimental and Theoretical Physics Letters*, 26:178, August 1977.

[139] M. E. Gertsenshtein and V. I. Pustovoit. On the detection of low-frequency gravitational waves. *Soviet Journal of Experimental and Theoretical Physics*, 16:433, January 1963.

[140] N. K. Glendenning and F. Weber. Nuclear solid crust on rotating strange quark stars. *The Astrophysical Journal*, 400:647–658, December 1992.

[141] S. K. Godunov. Difference method of computation of shock waves. *Uspekhi Matematicheskikh Nauk*, 12:176–177, 1957.

[142] S. K. Godunov. A difference method for numerical calculation of discontinuous solutions of fluid dynamics equations. *Matematičeskij Sbornik*, 47:271–306, 1959.

[143] S. K. Godunov and V. S. Ryabenkiy. *Difference Schemes. An Introduction to the Underlining Theory.* North Holland, 1987.

[144] P. Goldreich and Gerald Schubert. Differential rotation in stars. *The Astrophysical Journal*, 150:571, November 1967.

[145] J. Goodman. Are gamma-ray bursts optically thick? *The Astrophysical Journal*, 308:L47–L50, September 1986.

[146] J. Gorosabel, *et al.* Simultaneous polarization monitoring of supernovae SN 2008D/XT 080109 and SN 2007uy: Isolating geometry from dust. *Astronomy and Astrophysics*, 522:A14, November 2010.

[147] E. Göğüş, *et al.* Temporal and spectral characteristics of short bursts from the soft gamma repeaters 1806-20 and 1900+14. *The Astrophysical Journal*, 558:228–236, September 2001.

[148] O. M. Grimsrud and I. Wasserman. Non-equilibrium effects in steady relativistic ê + ê-gamma winds. *Monthly Notices of the Royal Astronomical Society*, 300:1158–1180, November 1998.

[149] L. P. Grishchuk, *et al.* Reviews of topical problems: Gravitational wave astronomy: In anticipation of first sources to be detected. *Physics Uspekhi*, 44(1):R01, January 2001.

[150] I. Hachisu. A versatile method for obtaining structures of rapidly rotating stars. *The Astrophysical Journal Supplement Series*, 61:479, July 1986.

[151] I. Hachisu, Y. Eriguchi, and K. Nomoto. Fate of merging double white dwarfs. *The Astrophysical Journal*, 308:161, September 1986.

[152] I. Hachisu. A versatile method for obtaining structures of rapidly rotating stars. II. Three-dimensional self-consistent field method. *The Astrophysical Journal Supplement Series*, 62:461, November 1986.

[153] I. Hachisu, Y. Eriguchi, and K. Nomoto. Fate of merging double white dwarfs. II: Numerical method. *The Astrophysical Journal*, 311:214–225, December 1986.

[154] P. Haensel, A. Y. Potekhin, and D. G. Yakovlev. *Neutron Stars 1 : Equation of State and Structure*, Volume 326. New York: Springer, 2007.

[155] E. Hairer and G. Wanner. *Solving Ordinary Differential Equations II: Stiff and Differential-Algebraic Problems.* 2nd revised edition. Berlin: Springer, 1996.

[156] R. Hakim. Remarks on relativistic statistical mechanics. I. *Journal of Mathematical Physics*, 8:1315–1344, June 1967.

[157] G. Hall and J. M. Watt. *Modern Numerical Methods for Ordinary Differential Equations.* Oxford University Press, New York, 1976.

[158] M. Hamuy. Observed and physical properties of core-collapse supernovae. *The Astrophysical Journal*, 582(2):905–914, January 2003.

[159] H. Harleston and K. A. Holcomb. Numerical solution of the general relativistic Boltzmann equation for massive and massless particles. *The Astrophysical Journal*, 372:225–240, May 1991.

[160] L.Hernquist and J. P. Ostriker. A self-consistent field method for galactic dynamics. *The Astrophysical Journal*, 386:375, February 1992.

[161] K. Hirata, *et al.* Observation of a neutrino burst from the supernova SN1987A. *Physical Review Letters*, 58:1490–1493, April 1987.

[162] S. P. Hirshman, *et al.* BCYCLIC: A parallel block tridiagonal matrix cyclic solver. *Journal of Computational Physics*, 229:6392–6404, September 2010.

[163] R. A. Hulse and J. H. Taylor. Discovery of a pulsar in a binary system. *The Astrophysical Journal*, 195:L51–L53, January 1975.

[164] K. Hurley. The 4.5+/-0.5 soft gamma repeaters in review. In R. M. Kippen, R. S. Mallozzi, and G. J. Fishman (editors), *Gamma-Ray*

Bursts, 5th Huntsville Symposium, volume 526 of *American Institute of Physics Conference Series*, pp. 763–770, September 2000.

[165] U. Hwang and J. M. Laming. A Chandra X-ray survey of ejecta in the Cassiopeia a supernova remnant. *The Astrophysical Journal*, 746(2):130, February 2012.

[166] A. I. Ibrahim, *et al.* An unusual burst from soft gamma repeater SGR 1900+14: Comparisons with giant flares and implications for the magnetar model. *The Astrophysical Journal*, 558:237–252, September 2001.

[167] L. A. I'lkaeva and N. A. Popov. Hydrodynamic solutions for one-dimensional perturbations of an unstable detonation wave (in Russian). *Fiz. Goreniya Vzryva*, 3:20–26, May 1965. https://www.sibran.ru/upload/iblock/506/5061ed54c06214112bce245c92ae9c81.pdf.

[168] V. S. Imshennik. The probable scenario of a supernova explosion as a result of massive stellar core gravitational collapse. *Pis'ma v Astronomicheskii Zhurnal*, 18:489–504, January 1992.

[169] V. S. Imshennik. Physics of our days: Supernova explosions and historical chronology. *Physics Uspekhi*, 43:509–513, May 2000.

[170] V. S. Imshennik and V. M. Chechetkin. Thermodynamics under conditions of hot matter neutronization and the hydrodynamic stability of stars at late stages of evolution. *Astronomicheskij Zhurnal*, 47:929, October 1970.

[171] V. S. Imshennik and V. M. Chechetkin. Thermodynamics under conditions of hot matter neutronization and the hydrodynamic stability of stars at late stages of evolution. *Soviet Astronomy*, 14:747, April 1971.

[172] V. S. Imshennik, *et al.* Is the detonation burning in a degenerate carbon-oxygen core of a presupernova possible? *Astronomy Letters*, 25:206–214, April 1999.

[173] V. S. Imshennik and A. M. Khokhlov. Detonation-wave structure and nucleosynthesis in exploding carbon-oxygen cores and white dwarfs. *Soviet Astronomy Letters*, 10:262–265, August 1984.

[174] V. S. Imshennik and D. K. Nadezhin. Neutrino thermal conductivity in collapsing stars. *Soviet Journal of Experimental and Theoretical Physics*, 36:821, 1973.

[175] V. S. Imshennik and D. K. Nadezhin. Supernova 1987A in the large magellanic cloud: Observations and theory. *Astrophysics and Space Physics Review*, 8:1–147, September 1989.

[176] V. S. Imshennik and D. K. Nadyozhin. SN 1987A and rotating neutron star formation. *Soviet Astronomy Letters*, 18:79, March 1992.

[177] L. N. Ivanova and V. M. Chechetkin. The collapse of degenerate iron stellar cores and the model of a supernova. *Soviet Astronomy*, 25:584, October 1981.

[178] L. N. Ivanova, V. S. Imshennik, and V. M. Chechetkin. Pulsation regime of the thermonuclear explosion of a star's dense carbon core. *Astrophysics and Space Science*, 31:497–514, December 1974.

[179] L. N. Ivanova, V. S. Imshennik, and V. M. Chechetkin. Numerical calculations of the thermonuclear explosion of a degenerate carbon stellar core. *Soviet Astronomy*, 21:374–381, June 1977.

[180] L. N. Ivanova, V. S. Imshennik, and V. M. Chechetkin. Numerical calculations of the thermonuclear explosion of a degenerate carbon stellar core. *Astronomicheskij Zhurnal*, 54:661, June 1977.

[181] L. N. Ivanova, V. S. Imshennik, and V. M. Chechetkin. Physical statement of the problem of the thermonuclear explosion of a degenerate carbon stellar core. *Soviet Astronomy*, 21:197–204, April 1977.

[182] L. N. Ivanova, V. S. Imshennik, and V. M. Chechetkin. Physical statement of the problem of the thermonuclear explosion of a degenerate carbon stellar core. *Astronomicheskij Zhurnal*, 54:354, April 1977.

[183] L. N. Ivanova, V. S. Imshennik, and V. M. Chechetkin. Triggering of stellar collapse by thermonuclear burning in a degenerate carbon core. *Soviet Astronomy*, 21:571–581, October 1977.

[184] L. N. Ivanova, V. S. Imshennik, and V. M. Chechetkin. Triggering of stellar collapse by thermonuclear burning in a degenerate carbon core. *Astronomicheskij Zhurnal*, 54:1009, October 1977.

[185] L. N. Ivanova, V. S. Imshennik, and V. M. Chechetkin. Pulsation regime of the thermonuclear explosion of a star's dense carbon core (In Russian). *Astrophysics and Space Science*, 31:477, December 1974.

[186] S. Jackson. Rapidly rotating stars. V. The coupling of the henyey and the self-consistent methods. *The Astrophysical Journal*, 161:579, August 1970.

[187] R. A. James. The structure and stability of rotating gas masses. *The Astrophysical Journal*, 140:552, August 1964.

[188] H.-T. Janka. Explosion mechanisms of core-collapse supernovae. *Annual Review of Nuclear and Particle Science*, 62:407–451, November 2012.

[189] H.-T. Janka, *et al.* Theory of core-collapse supernovae. *Physics Reports*, 442:38–74, April 2007.

[190] I. M. Khalatnikov. *From the Atomic Bomb to the Landau Institute.* Springer-Verlag, 2012.

[191] A. M. Khokhlov. Stability of detonations in supernovae. *The Astrophysical Journal*, 419:200, December 1993.

[192] S. M. Kiselev, *et al.* The neutron-star equation of state and the two-nucleon nuclear interaction. *Soviet Journal of Nuclear Physics*, 29:708–712, 1979.

[193] A. V. Koldoba, E. V. Tarasova, and V. M. Chechetkin. Instability of the detonation wave in a thermonuclear supernova model. *Astronomy Letters*, 20(3):377–381, May 1994.

[194] A. N. Kolmogorov and S. V. Fomin. *Elements of the Theory of Functions and Functional Analysis.* Dover Publications, 1999.

[195] A. G. Kritsuk, *et al.* Simulating supersonic turbulence in magnetized molecular clouds. In *Journal of Physics Conference Series* volume 180. *Proceedings of the DOE/SciDAC 2009 Conference*, p. 012020, July 2009.

[196] M. W. Kutta. Beitrag zur näherungsweisen Integration totaler Differentialgleichungen. *Zeitschrift für Mathematik und Physik*, 46:435–453, 1901.

[197] O. A. Kuznetsov, *et al.* Orbital evolution of a binary neutron star and gravitational radiation. *Astronomy Reports*, 42(5):638–648, September 1998.

[198] L. D. Landau. To the stars theory. *Phys. Zs. Sowjet. (English and German)*, 1:285, December 1932.

[199] L. D. Landau and E. M. Lifshits. *Fluid Mechanics (Course of Theoretical Physics).* Pergamon, New York, 1987.

[200] L. D. Landau and E. M. Lifshitz. *Fluid Mechanics.* New York: Pergamon Press, 1959.

[201] L. D. Landau and E. M. Lifshitz. *The Classical Theory of Fields.* Oxford: Pergamon Press, 1975.

[202] L. D. Landau and E. M. Lifshitz. *Statistical Physics.* Pergamon Press, Oxford, 1980.

[203] K. R. Lang. *Astrophysical Formulae.* New York: Springer, 1999.

[204] K. R. Lang. *Astrophysical Formulae. A Compendium for the Physicist and Astrophysicist.* Berlin, Heidelberg, New York: Springer-Verlag, Springer Study Edition, 1980.

[205] N. Langer. Presupernova evolution of massive single and binary stars. *Annual Review of Astronomy and Astrophysics*, 50:107–164, September 2012.

[206] J. M. Lattimer and D. F. Swesty. A generalized equation of state for hot, dense matter. *Nuclear Physics A*, 535(2):331–376, December 1991.

[207] J. M. Lattimer and F. D. Swesty. Lattimer-swesty equation of state code. Astrophysics Source Code Library, record ascl:1202.011, February 2012.

[208] J. M. LeBlanc and J. R. Wilson. A numerical example of the collapse of a rotating magnetized star. *The Astrophysical Journal*, 161:541, August 1970.

[209] P. Ledoux. Stellar models with convection and with discontinuity of the mean molecular weight. *The Astrophysical Journal*, 105:305, March 1947.

[210] E. J. Lentz, *et al.* On the requirements for realistic modeling of neutrino transport in simulations of core-collapse supernovae. *The Astrophysical Journal*, 747:73, March 2012.

[211] M. Liebendörfer, *et al.* A finite difference representation of neutrino radiation hydrodynamics in spherically symmetric general relativistic spacetime. *The Astrophysical Journal Supplement Series*, 150:263–316, January 2004.

[212] Y. T. Liu, *et al.* General relativistic simulations of magnetized binary neutron star mergers. *Physical Review D*, 78(2):024012, July 2008.

[213] C. Y. Loh and P. C. E. Jorgenson. Multi-dimensional dissipation for cure of pathological behaviors of upwind scheme. *Journal of Computational Physics*, 228(5):1343–1346, March 2009.

[214] K. Maeda, *et al.* Nucleosynthesis in two-dimensional delayed detonation models of type Ia Supernova Explosions. *The Astrophysical Journal*, 712(1):624–638, March 2010.

[215] K. Maeda, *et al.* Explosive nucleosynthesis in aspherical hypernova explosions and late-time spectra of SN 1998bw. *The Astrophysical Journal*, 565(1):405–412, January 2002.

[216] J. W. K. Mark. Rapidly rotating stars. III. Massive main-sequence stars. *The Astrophysical Journal*, 154:627, November 1968.

[217] L. A. Marschall. *The Supernova Story.* New York: Plenum Press, 1988.

[218] P. A. Mazzali, *et al.* Models for the type Ic hypernova SN 2003lw associated with GRB 031203. *The Astrophysical Journal*, 645(2):1323–1330, July 2006.

[219] D. L. Meier, *et al.* Magnetohydrodynamic phenomena in collapsing stellar cores. *The Astrophysical Journal*, 204:869–878, March 1976.

[220] A. Mezzacappa and S. W. Bruenn. A numerical method for solving the neutrino Boltzmann equation coupled to spherically symmetric stellar core collapse. *The Astrophysical Journal*, 405:669–684, March 1993.

[221] A. Mezzacappa and S. W. Bruenn. Stellar core collapse: A Boltzmann treatment of neutrino-electron scattering. *The Astrophysical Journal*, 410:740–760, June 1993.

[222] A. Mezzacappa and S. W. Bruenn. Type II supernovae and Boltzmann neutrino transport: The infall phase. *The Astrophysical Journal*, 405:637–668, March 1993.

[223] A. Mezzacappa, *et al.* Simulation of the spherically symmetric stellar core collapse, bounce, and postbounce evolution of a star of 13 solar masses with Boltzmann neutrino transport, and its implications for the supernova mechanism. *Physical Review Letters*, 86:1935–1938, March 2001.

[224] D. Mihalas and B. W. Mihalas. *Foundations of Radiation Hydrodynamics.* Oxford University Press, New York, 1984.

[225] G. H. Miller and E. G. Puckett. A high-order Godunov method for multiple condensed phases. *Journal of Computational Physics*, 128:134–164, 1996.

[226] A. Mirizzi, *et al.* Supernova neutrinos: Production, oscillations and detection. *Nuovo Cimento Rivista Serie*, 39:1–112, 2016.

[227] C. W. Misner, K. S. Thorne, and J. A. Wheeler. *Gravitation.* San Francisco: W.H. Freeman and Company, 1973.

[228] S. G. Moiseenko, G. S. Bisnovatyi-Kogan, and N. V. Ardeljan. Magnetorotational processes in core collapse supernovae. In Sandip K. Chakrabarti, Alexander I. Zhuk, and Gennady S. Bisnovatyi-Kogan (editors), *American Institute of Physics Conference Series*, volume 1206. *Proceedings of the 4th Gamow International Conference on Astrophysics and Cosmology After Gamow* and the 9th Gamow Summer School "Astronomy and Beyond: Astrophysics, Cosmology, Radio Astronomy, High Energy Physics and Astrobiology", pp. 282–292, January 2010.

[229] P. Mösta, *et al.* Magnetorotational core-collapse supernovae in three dimensions. *The Astrophysical Journal*, 785:L29, April 2014.

[230] E. Mueller and Y. Eriguchi. Equilibrium models of differentially rotating, completely catalyzed, zero-temperature configurations with central densities intermediate to white dwarf and neutron star densities. *Astronomy and Astrophysics*, 152(2):325–335, November 1985.

[231] B. Müller, H.-T. Janka, and H. Dimmelmeier. A new multi-dimensional general relativistic neutrino hydrodynamic code for core-collapse supernovae. I. Method and code tests in spherical symmetry. *The Astrophysical Journal Supplement Series*, 189:104–133, July 2010.

[232] J. W. Murphy and C. Meakin. A global turbulence model for neutrino-driven convection in core-collapse supernovae. *The Astrophysical Journal*, 742:74, December 2011.

[233] E. S. Myra and A. Burrows. Neutrinos from type II supernovae: The first 100 milliseconds. *The Astrophysical Journal*, 364:222–231, November 1990.

[234] D. K. Nadezhin. The collapse of iron-oxygen stars: Physical and mathematical formulation of the problem and computational method. *Astrophysics and Space Science*, 49:399–425, July 1977.

[235] D. K. Nadezhin. The neutrino radiation for a hot neutron star formation and the envelope outburst problem. *Astrophysics and Space Science*, 53:131–153, January 1978.

[236] D. K. Nadezhin and V. M. Chechetkin. Neutrino radiation by the URCA process at high temperatures. *Astronomicheskij Zhurnal*, 46:270, 1969.

[237] H. Nagakura. Retrieval of energy spectra for all flavours of neutrinos from core-collapse supernova with multiple detectors. *MNRAS*, 500(1):319–332, January 2021.

[238] H. Nagakura, A. Burrows, and D. Vartanyan. Supernova neutrino signals based on long-term axisymmetric simulations. *MNRAS*, 506(1):1462–1479, September 2021.

[239] S. Nagataki. Influence of axisymmetrically deformed explosions in type II supernovae on the reproduction of the solar system abundances. *The Astrophysical Journal*, 511(1):341–350, January 1999.

[240] S. Nagataki, *et al.* Explosive nucleosynthesis in axisymmetrically deformed type II supernovae. *The Astrophysical Journal*, 486(2):1026–1035, September 1997.

[241] S. Nagataki, A. Mizuta, and K. Sato. Explosive nucleosynthesis in GRB jets accompanied by hypernovae. *The Astrophysical Journal*, 647(2):1255–1268, August 2006.

[242] T. Nakamura, *et al.* Light curve and spectral models for the hypernova SN 1998BW associated with GRB 980425. *The Astrophysical Journal*, 550(2):991–999, April 2001.

[243] J. C. Niemeyer and W. Hillebrandt. Microscopic instabilities of nuclear flames in type IA supernovae. *The Astrophysical Journal*, 452:779, October 1995.

[244] J. C. Niemeyer and W. Hillebrandt. Turbulent nuclear flames in type IA supernovae. *The Astrophysical Journal*, 452:769, October 1995.

[245] K. Nomoto and M. Hashimoto. Presupernova evolution of massive stars. *Physics Reports*, 163:13–36, 1988.

[246] E. O'Connor and C. D. Ott. A new open-source code for spherically symmetric stellar collapse to neutron stars and black holes. *Classical and Quantum Gravity*, 27(11):114103, June 2010.

[247] T. Ohkubo, *et al.* Core-collapse very massive stars: Evolution, explosion, and nucleosynthesis of population III 500–1000 M_{solar} stars. *The Astrophysical Journal*, 645(2):1352–1372, July 2006.

[248] O. Østerby and Z. Zlatev. *Direct Methods for Sparse Matrixes. Lecture notes in Computer Science.*, Vol. 157, Springer, Berlin, Heidelberg, New York, Tokio, 1983.

[249] J. P. Ostriker and J. E. Gunn. Do pulsars make supernovae? *The Astrophysical Journal*, 164:L95, March 1971.

[250] J. P. Ostriker and J. W. K. Mark. Rapidly rotating stars. I. The self-consistent-field method. *The Astrophysical Journal*, 151:1075–1088, March 1968.

[251] J. P. Ostriker and P. Bodenheimer. Rapidly rotating stars. II. Massive white dwarfs. *The Astrophysical Journal*, 151:1089, March 1968.

[252] B. Paczynski. Gamma-ray bursters at cosmological distances. *The Astrophysical Journal*, 308:L43–L46, September 1986.

[253] B. Paczynski. Super-Eddington winds from neutron stars. *The Astrophysical Journal*, 363:218–226, November 1990.

[254] F. Patat, *et al.* The metamorphosis of SN 1998bw. *The Astrophysical Journal*, 555(2):900–917, July 2001.

[255] G. G. Pavlov and D. G. Yakovlev. Meridional circulation caused by stellar rotation. *Soviet Astronomy*, 22:595–602, October 1978.

[256] M. Pelanti and K.-M. Shyue. A mixture-energy-consistent six-equation two-phase numerical model for fluids with interfaces, cavitation and evaporation waves. *Journal of Computational Physics*, 259:331–357, February 2014.

[257] M. V. Popov. Piecewise parabolic method on a local stencil in cylindrical coordinates for fluid dynamics simulations. *Computational Mathematics and Mathematical Physics*, 52(8):1186–1201, August 2012.

[258] M. V. Popov, *et al.* Aspherical nucleosynthesis in a core-collapse supernova with 25 $M_{\odot}$ standard progenitor. *The Astrophysical Journal*, 783(1):43, March 2014.

[259] M. V. Popov and S. D. Ustyugov. Piecewise parabolic method on local stencil for gasdynamic simulations. *Computational Mathematics and Mathematical Physics*, 47(12):1970–1989, December 2007.

[260] D. E. Potter. *Computational Physics.* John Wiley & Sons Ltd., 1973.

[261] V. V. Pukhnachev. The stability of Chapman-Jouguet detonations. *Soviet Physics Doklady*, 8:338, January 1963.

[262] J. J. Quirk. A contribution to the great Riemann solver debate. *International Journal for Numerical Methods in Fluids*, 18(6):555–574, March 1994.

[263] D. Radice, *et al.* Neutrino-driven convection in core-collapse supernovae: High-resolution simulations. *The Astrophysical Journal*, 820:76, March 2016.

[264] P. K. Raschewski. *Riemannsche Geometrie und Tensoranalysis.* Frankfurt am Main: Verlag Harri Deutsch, 2. unveränd. Aufl. Edition, 1995.

[265] T. Rauscher and F.-K. Thielemann. Astrophysical reaction rates from statistical model calculations. *Atomic Data and Nuclear Data Tables*, 75(1–2):1–351, May 2000.

[266] M. Rees, R. Ruffini, and J. A. Wheeler. *Black Holes, Gravitational Waves and Cosmology: An Introduction to Current Research.* New York: Gordon and Breach, Science Publishers, Inc., 1974.

[267] R. Richtmeyer and K. Morton. *Difference Methods for Initial Value Problems.* Wiley, New York, 1967.

[268] P. L. Roe. Approximate Riemann solvers, parameter vectors, and difference schemes. *Journal of Computational Physics*, 135:250–258, August 1997.

[269] B. L. Rozdestvenskii and N. N. Janenko. *Systems of Quasilinear Equations and Their Applications to Gas Dynamics.* American Mathematical Society, 1983.

[270] M. Ruffert, H. T. Janka, and G. Schaefer. Coalescing neutron stars: A step towards physical models. I. Hydrodynamic evolution and gravitational-wave emission. *Astronomy and Astrophysics*, 311:532–566, July 1996.

[271] M. Ruffert, *et al.* Coalescing neutron stars: A step towards physical models. II. Neutrino emission, neutron tori, and gamma-ray bursts. *Astronomy and Astrophysics*, 319:122–153, March 1997.

[272] M. Ruffert and H. Th. Janka. Coalescing neutron stars: A step towards physical models. III. Improved numerics and different neutron star masses and spins. *Astronomy and Astrophysics*, 380:544–577, December 2001.

[273] O. G. Ryazhskaya. Neutrinos from stellar core collapses: Present status of experiments. *Physics Uspekhi*, 49:1017–1027, October 2006.

[274] W. E. Schiesser. *The Numerical Method of Lines.* Academic Press, 1991.

[275] L. I. Sedov. Propagation of strong shock waves. *Journal of Applied Mathematics and Mechanics*, 10:241–250, January 1946.

[276] I. R. Seitenzahl, *et al.* Nucleosynthesis in thermonuclear supernovae with tracers: Convergence and variable mass particles. *Monthly Notices of the Royal Astronomical Society*, 407(4):2297–2304, October 2010.

[277] A. S. Sengupta, LIGO Scientific Collaboration, and Virgo Collaboration. LIGO-Virgo searches for gravitational waves from coalescing binaries: A status update. In *Journal of Physics Conference Series*, volume 228, p. 012002, May 2010. *Proceedings of the 8th E. Amaldi Conference on Gravitational Waves (Amaldi8)*, New York, 2009.
[278] V. D. Shafranov. The structure of shock waves in a plasma. *Sov. Phys. JETP*, 5:1183–1188, 1957.
[279] S. L. Shapiro and S. A. Teukolsky. *Black Holes, White Dwarfs, and Neutron Stars: The Physics of Compact Objects.* A Wiley-Interscience Publication, New York: Wiley, 1983.
[280] M. Shibata, *et al.* Magnetized hypermassive neutron-star collapse: A central engine for short gamma-ray bursts. *Physical Review Letters*, 96(3):031102, January 2006.
[281] M. A. Skinner, *et al.* FORNAX: A flexible code for multiphysics astrophysical simulations. *The Astrophysical Journal Supplement Series*, 241(1):7, March 2019.
[282] S. J. Smartt. Progenitors of core-collapse supernovae. *Annual Review of Astronomy and Astrophysics*, 47:63–106, September 2009.
[283] S. J. Smartt, *et al.* The death of massive stars: I. Observational constraints on the progenitors of Type II-P supernovae. *Monthly Notices of the Royal Astronomical Society*, 395(3):1409–1437, May 2009.
[284] G. D. Smith. *Numerical Solution of Partial Differential Equations: Finite Difference Methods.* Clarendon Press, Oxford, 1985.
[285] A. M. Soderberg, *et al.* An extremely luminous X-ray outburst at the birth of a supernova. *Nature*, 453(7194):469–474, May 2008.
[286] R. Stoeckly. Polytropic models with fast, non-uniform rotation. *The Astrophysical Journal*, 142:208–228, July 1965.
[287] V. M. Suslin, *et al.* Simulation of neutrino transport by large-scale convective instability in a proto-neutron star. *Astronomy Reports*, 45:241–247, March 2001.
[288] F. D. Swesty and E. S. Myra. A numerical algorithm for modeling multigroup neutrino-radiation hydrodynamics in two spatial dimensions. *The Astrophysical Journal Supplement Series*, 181:1–52, March 2009.
[289] I. Tamborra, *et al.* Self-sustained asymmetry of Lepton-number Emission: A new phenomenon during the supernova shock-accretion phase in three dimensions. *The Astrophysical Journal*, 792:96, September 2014.
[290] M. Tanaka, *et al.* Nebular phase observations of the type Ib supernova 2008D/X-ray transient 080109: Side-viewed bipolar explosion. *The Astrophysical Journal*, 700(2):1680–1685, August 2009.

[291] J.-L. Tassoul. *Theory of Rotating Stars.* Princeton Series in Astrophysics, Princeton: Princeton University Press, 1978.

[292] J. Thijssen. *Computational Physics.* Cambridge University Press, March 2007.

[293] H. C. Thomas. Consequences of mass transfer in close binary systems. *Annual Review of Astronomy Astrophysics*, 15:127–151, 1977.

[294] C. Thompson and R. C. Duncan. The soft gamma repeaters as very strongly magnetized neutron stars - I. Radiative mechanism for outbursts. *Monthly Notices of the Royal Astronomical Society*, 275:255–300, July 1995.

[295] A. N. Tikhonovl and A. A. Samarskii. *Equations of Mathematical Physics.* Dover Publications, 1990.

[296] F. X. Timmes and S. E. Woosley. The conductive propagation of nuclear flames. I. degenerate C + O and O + NE + MG white dwarfs. *The Astrophysical Journal*, 396:649, September 1992.

[297] C. Travaglio, *et al.* Nucleosynthesis in multi-dimensional SN Ia explosions. *Astronomy and Astrophysics*, 425:1029–1040, October 2004.

[298] C. Travaglio, K. Kifonidis, and E. Müller. Nucleosynthesis in multi-dimensional simulations of SNII. *New Astronomy Reviews*, 48 (1–4):25–30, February 2004.

[299] J. Trumper. X-ray observations of neutron stars and the equation of state at supra-nuclear densities. *New Astronomy Reviews*, 54:122–127, March 2010.

[300] D. L. Tubbs and D. N. Schramm. Neutrino opacities at high temperatures and densities. *The Astrophysical Journal*, 201:467–488, October 1975.

[301] K. Uryu and Y. Eriguchi. Structures of rapidly rotating baroclinic stars: Part one: A numerical method for the angular velocity distribution. *Monthly Notices of the Royal Astronomical Society*, 269:24, July 1994.

[302] V. V. Usov. Bare quark matter surfaces of strange stars and e^+e^- emission. *Physical Review Letters*, 80:230–233, January 1998.

[303] V. V. Usov. Strange star heating events as a model for giant flares of soft-gamma-ray repeaters. *Physical Review Letters*, 87(2):021101, July 2001.

[304] V. V. Usov. The response of bare strange stars to the energy input onto their surfaces. *The Astrophysical Journal*, 559:L135–L138, October 2001.

[305] V. V. Usov. Thermal emission from bare quark matter surfaces of hot strange stars. *The Astrophysical Journal*, 550:L179–L182, April 2001.

[306] V. V. Usov, T. Harko, and K. S. Cheng. Structure of the electrospheres of bare strange stars. *The Astrophysical Journal*, 620:915–921, February 2005.

[307] S. D. Ustyugov and V. M. Chechetkin. Supernovae explosions in the presence of large-scale convective instability in a rotating protoneutron star. *Astronomy Reports*, 43(11):718–726, November 1999.

[308] S. D. Ustyugov, *et al.* Piecewise parabolic method on a local stencil for magnetized supersonic turbulence simulation. *Journal of Computational Physics*, 228(20):7614–7633, November 2009.

[309] G. V. Vereshchagin and A. G. Aksenov. *Relativistic Kinetic Theory.* Cambridge University Press, 2017.

[310] N. Ya. Vilenkin. *Special Functions and the Theory of Group Representations.* Providence: American Mathematical Society. 1968.

[311] V. S. Vladimirov. Equations of mathematical physics. In A. Jeffrey (editor), *Pure and Applied Mathematics*, 3(vi):418. New York: Marcel Dekker, Inc., 1971.

[312] J. von Neumann. *Mathematical Foundations of Quantum Mechanics.* Princeton University Press, 1996.

[313] T. A. Weaver, G. B. Zimmerman, and S. E. Woosley. Presupernova evolution of massive stars. *The Astrophysical Journal*, 225:1021–1029, November 1978.

[314] F. Weber. Strange quark matter and compact stars. *Progress in Particle and Nuclear Physics*, 54:193–288, March 2005.

[315] J. C. Wheeler, D. L. Meier, and J. R. Wilson. Asymmetric supernovae from magnetocentrifugal jets. *The Astrophysical Journal*, 568:807–819, April 2002.

[316] E. Witten. Cosmic separation of phases. *Physical Review D*, 30:272–285, July 1984.

[317] S. Wolfram. *The Mathematica Book.* Cambridge University Press, 1996.

[318] A. Wongwathanarat, E. Müller, and H.-T. Janka. Three-dimensional simulations of core-collapse supernovae: From shock revival to shock breakout. *Astronomy and Astrophysics*, 577:A48, May 2015.

[319] S. E. Woosley, A. Heger, and T. A. Weaver. The evolution and explosion of massive stars. *Reviews of Modern Physics*, 74(4):1015–1071, November 2002.

[320] S. E. Woosley and T. A. Weaver. The evolution and explosion of massive stars. II. Explosive hydrodynamics and nucleosynthesis. *The Astrophysical Journal Supplement Series*, 101:181, November 1995.

[321] R. X. Xu, G. J. Qiao, and B. Zhang. Are pulsars bare strange stars? In M. Kramer, N. Wex, and R. Wielebinski, editors, *IAU Colloq. 177:*

Pulsar Astronomy — 2000 and Beyond, volume 202 of *Astronomical Society of the Pacific Conference Series*, p. 665, 2000.

[322] D. G. Yakovlev, *et al.* Lev Landau and the concept of neutron stars. *Physics Uspekhi*, 56(3):289–295, March 2013.

[323] B. G. Zakharov. Photon emission from bare quark stars. *Soviet Journal of Experimental and Theoretical Physics*, 112:63–76, January 2011.

[324] G. I. Zatsepin. Probability of determining the upper limit of the neutrino mass from the time of flight. *Soviet Journal of Experimental and Theoretical Physics Letters*, 8:205, September 1968.

[325] I. B. Zeldovich and I. D. Novikov. *Relativistic Astrophysics. Volume 2: The Structure and Evolution of the Universe.* (Revised and enlarged edition) Chicago, IL: University of Chicago Press, 1983.

[326] Ya B. Zeldovich and Yu. P. Raizer. *Elements of Gasdynamics and the Classical Theory of Shock Waves.* New York: Academic Press, 1966.

[327] V. T. Zhukov, A. V. Zabrodin, and O. B. Feodoritova. A method for solving two-dimensional equations of heat-conducting gas dynamics in domains of complex configurations. *Zhurnal Vychislitelnoi Matematiki i Matematicheskoi Fiziki*, 33:1240–1250, August 1993.

[328] N. V. Zmitrenko, *et al.* Neutrino mechanism of propagation of thermonuclear burning in degenerate cores of stars. *Soviet Journal of Experimental and Theoretical Physics*, 48:589, October 1978.

Index

I

J

K

L

M

N

O

P

Q

R

S

Printed in the United States
by Baker & Taylor Publisher Services